WEATHER EYE— THE FINAL YEAR

WEATHER EYE—THE FINAL YEAR

An anthology, compiled by Anne McWilliams

BRENDAN McWILLIAMS ~

Gill & Macmillan

Gill & Macmillan Ltd
Hume Avenue, Park West, Dublin 12
with associated companies throughout the world
www.gillmacmillan.ie

978 07171 4608 6

Index compiled by Cover to Cover
Typography design by Make Communication
Print origination by Carole Lynch
Printed in the UK by CPI Mackays, Chatham

This book is typeset in Linotype Minion and
Neue Helvetica.

The paper used in this book comes from the wood pulp of
managed forests. For every tree felled, at least one tree is
planted, thereby renewing natural resources.

A CIP catalogue record for this book is available
from the British Library.

5 4 3 2 1

Dedication

Once again, dearest Brendan
this is for you

ACKNOWLEDGMENTS

I would like to thank my literary agent, Jonathan Williams, whose help and advice throughout the whole publishing process has been invaluable. The team at Gill & Macmillan also deserve my gratitude—they have been a delight to work with. Lastly, I am very grateful to Peter Lynch, Professor of Meteorology at UCD, for looking over this and the previous *Book of Weather Eye* for any inconsistencies and gremlins which may have crept into the scientific text.

On a more personal note, I am indebted to my family and friends, each and every one of them, for their encouragement, patience and support since Brendan died. It is true to say that this book could not have been compiled without that support and I will be forever grateful to them.

FOREWORD

A Weather Eye on Weather Eye

Mark Twain once opined that 'Everybody talks about the weather but nobody does anything about it'. If, alas, nothing can be done to change the elements, the next best thing, surely, is to write about them. In this, Brendan McWilliams excelled. His daily *Weather Eye* column, which ran in the *Irish Times* for almost two decades, communicated articulately the science of meteorology and much more besides. Woven delicately with threads of weather lore, history, literature, art, music, astronomy, politics and the environment, it was as diverse as it was informative and, often, the first column to which readers turned each morning.

This book contains 100 or more selected articles from the final year of *Weather Eye*. They are the work of a man who, although easygoing in nature, possessed great depth, intelligence, creativity and perspective. There was little he would not tolerate in this world except, perhaps, intolerance itself and people who took themselves too seriously. His curiosity is reflected in every article, each the result of fastidious research, the end point of much ferreting out of information by a man who regarded himself as a kind of literary 'magpie', ever on the lookout for serendipitous nuggets of knowledge.

Where else might one chance upon an explanation of 'St Maury's Wind' or the unfortunate manner in which St Placid fell into a lake? Indeed, it is in another biblically themed offering that Dad's talent for attention-grabbing opening lines is displayed. Before launching into various scientific explanations for the parting of the Red Sea described in the Book of Exodus, he remarks, 'Sometimes, while reading their Bibles, meteorologists try to imagine the means by which the Almighty might have engineered His multitude of miracles.' Dad's humour was, of course, self-directed; there were few people less likely to be found reading the Bible for religious purposes.

In another inspired opening line, he refers to 'Weather Eye and its entourage, currently on tour in northern Britain'. The occasion was a well-earned holiday for my parents—just the two of them—in Yorkshire and Scotland. My mother, of course, was the aforementioned 'entourage', a term of endearment that got my father into considerable trouble indeed. But it was the humorous hook that often drew the reader in, a kind of avenue that paved the way for an explanation of some scientific fact or point of view.

His closing lines weren't bad either. The end of his article about Michael Crichton's sceptical novel on global warming, *State of Fear*, is a case in point. In this piece, *Weather Eye* despairs at the manner in which genuine but highly selective references are used to paint a picture of climate change far removed from the current consensus. Indeed, as Dad recounts of the book's central characters, 'any unfortunate passer-by who says anything about global warming is treated to a rude and arrogant tirade of several pages to persuade him, or her, of the total error of such views.' But it is the last line that leaves the reader in no doubt about the views of the columnist who, 'seeing that one of the characters was a Mr Balder, was disappointed not to find another by the name of Dash'.

Dad's unembellished environmental views were evident in many of his articles, not least in that entitled *Food for Climatic Thought*. Citing the *Stern Review on the Economics of Climate Change*, he distils for the reader, the report's predictions down to their basic components: sudden shifts in regional weather patterns; disease and malnutrition on a grand scale; millions displaced by rising sea levels; violent conflicts in vulnerable parts of the world; and a disruption to economic activity comparable to the World Wars and economic depressions of the twentieth century. When asked (as he occasionally was) whether he thought it was possible that humankind would reach a consensus and halt the kind of activities that appear to drive unprecedented climate change, Dad's response was usually quite simply 'No.'

It was during the year in which these articles were first published that Dad began his weekly interviews on RTÉ's *Today with Pat Kenny*. Having travelled by train from Wexford each Tuesday, he would appear on the show shortly before noon and then take a taxi to a local pub where he would have lunch with his Uncle Oliver, his friend Eamon, and me. Following this weekly ritual, I would drive him back to the station again in time to catch the afternoon train. The radio, of course, provided him with a fountain of ideas. Ever grateful for the letters and emails he received daily from both readers and listeners, he often used their suggestions in subsequent articles.

One such example was from a listener who phoned the radio show to ask the question: which is the commonest cloud in Irish skies? An article entitled *A Catalogue of Common Clouds* appeared in the *Irish Times* shortly after, on 20 September 2007. Ironically, it was at this time—almost to the day—that Dad had to discontinue his weekly sojourns on radio because of ill-health. Indeed, in a sense, this book has no natural end. Poignantly, it simply stops on 3 October 2007 without any warning of the tragedy that was yet to come.

Throughout his success as a writer, Dad was a modest man, a true gentleman who lived, as he wrote, with great warmth and insight. Some of the best examples of his work are to be found within this volume, where he combines his subtle and mischievous sense of humour with his thirst for knowledge and a desire to impart information of interest and importance. Here I have highlighted just a few examples; no doubt, everyone will have their favourite. Enjoy.

Stephen McWilliams
March 2009

MARS MADE MARVELLOUS

27 *October* 2006 ∿

In 1976, NASA scientists studying images of the Martian surface taken by the Viking 1 Orbiter noticed that one of the
features captured by the camera was distinctive. They were
suitably amused, and published the image shortly afterwards with
a press release in which they drew attention to a formation
that 'resembles a human head'. Thus began the 'face on Mars'
controversy which has continued for thirty years between main-
stream astronomers and the slightly gaga wing of that activity.

The argument is reminiscent of a similar controversy near the
end of the nineteenth century. In 1877, an Italian astronomer
called Giovanni Schiaparelli detected on Mars a network of
straight lines, which he called *canali*, the Italian word for channels.
Others, however, claimed these were irrigation canals dug by an
intelligent race of Martians to transport water from the melting
polar icecaps to the arid lower latitudes. The notion gained wide
acceptance at the time—it was even suggested that a message for
our Martian cousins might be drawn on the Sahara—but time
and better telescopes laid Schiaparelli's Martian ghosts to rest.

The area now in question is the Cydonia region of the planet,
and the observed formation lies close to an escarpment which
separates heavily cratered highlands from a low-lying plain. Its
elements are similar to pyramids in shape, and are arranged—
presumably fortuitously—in such a way that they resemble a great
human face staring into the Martian sky from the surface of the
planet.

There are those who maintain that this 'face' cannot be
explained by natural processes, and that what we see may be the

remains of a once thriving Martian city. Others believe it to be an artefact deliberately left there centuries ago by intelligent beings of extraterrestrial origin who have since moved on; it has been suggested that the face is similar to the Egyptian Sphinx at Giza, and even that it bears a close resemblance to the face on the Turin Shroud.

NASA, needless to say, is having none of this. Its scientists assert that the formations are 'a combination of lighting, shadowing and natural erosion', and point out that if you look at any pattern for long enough, you can see a face, a tree, or other familiar shape; they have even, uncharacteristically, quoted Antony in Shakespeare's *Antony and Cleopatra* as they make their point:

> *Sometimes we see a cloud that's dragonish;*
> *A vapour sometime like a bear or lion,*
> *A tower'd citadel, a pendant rock,*
> *A forked mountain, or blue promontory*
> *With trees upon't, that nod unto the world*
> *And mock our eyes. . . .*

Within the last few days the European Space Agency has released a new high-resolution 3-dimensional animated film-loop of the face on Mars, so if you have broadband Internet you can check it out at www.esa.int/SPECIALS/Mars_Express/SEMINCO7BTE _0.html and draw your own conclusions.

WHITHER CAME THE MOON?

4 November 2006 ∿

With clear skies these nights, and Hunters' Moon near full, you may now and then, like John Milton,

. . . walk unseen
On the dry smooth-shaven green,
And behold the wandering Moon
Riding near her highest noon.

And indeed you may ask yourself a question that thinking sorts have asked themselves for centuries: where did it come from, this great celestial sphere that dominates our nightly skies at monthly intervals?

The first known attempt to explain the Moon's origins was in the fifth century BC when the Greek philosopher Anaxagoras, having inspected a meteorite that had fallen from the sky, speculated that all celestial bodies were glowing 'stone stars' that had been flung out from Earth. As it happens, and as we shall see later, he got it nearly right in the particular case of the Moon, but astronomy textbooks only a generation ago were more unsure.

One theory dating back to 1879 was suggested by George Howard Darwin, who was determined not to be outdone by his famous father, Charles, when it came to proposing new and adventurous scientific theories. He was of the view that the Moon was a piece of Earth which had broken loose many millions of years ago, leaving the Pacific Ocean as the deep scar of the separation. It was a popular theory in its day, although it commands little credence nowadays.

The reclusive mathematician Edouard Roche, on the other hand, was inclined to think that the Moon was formed independently, out of the same cosmic dust that produced the Earth itself. Indeed some astronomers, like the eccentric T.J.J. See, maintained that the Moon was once a planet in its own right, travelling proudly like the Earth in an elliptical orbit around the Sun. It may have circled in this way for many billions of years, they suggested, before a rare combination of the orbits of itself, Earth, and perhaps Venus caused it to be 'captured' by our planet, making it a prisoner forever annexed to a neighbouring Great Power, condemned *ad infinitum* to revolve around us at monotonous four-weekly intervals.

But when fragments of the Moon were brought back to Earth and analysed following the Apollo landings, it was found that none of these theories fitted all the facts. Astronomers in the 1970s then devised what might seem like an even more bizarre scenario. According to the 'Big Splash' theory, billions of years ago a Moonless Earth was in collision with a Mars-sized planet; this giant impact melted the upper layers of the Earth's crust to form an ocean of molten rock some 700 miles in depth, and the large splash from this molten ocean sent a cloud of incandescent fragments and hot gases surging back into space. The more volatile substances comprising this debris, the theory goes, vaporised almost immediately to nothingness, but the remainder eventually coalesced to form the Moon.

CHARLES DICKENS'S NOUVELLE CUISINE

6 November 2006 ∽

I n the celebrated case of *Bardell* v. *Pickwick*, the widow Bardell wins substantial damages for breach of promise of marriage from the unfortunate Mr Pickwick. The most damning piece of evidence is read to the court by the plaintiff's counsel, Mr Serjeant Buzfuz. 'Garraway's, twelve o'clock,' it reads, 'Chops and Tomata sauce. Yours, Pickwick.'

Now one of the more interesting features of this incriminating communication is the relevant novelty of 'tomata' at the time. *The Posthumous Papers of the Pickwick Club*, although written in 1836, are required to be eponymously posthumous to the eponymous association, and are therefore set in the middle 1820s, at a time when the tomato had only relatively recently begun to figure prominently in everyday cuisine.

The wild plant was native to Peru, to a region between latitudes 5° and 15° south of the equator. From there it had spread north-wards to Mexico, where it was domesticated by the Incas and given the name *xicotomatl*, before being brought to Europe by the Spaniards. It grew easily in the Mediterranean climate from around 1540 onwards, but for some considerable time Europeans were reluctant to think of it as edible; as a member of the deadly nightshade family, it was assumed to be poisonous and was there-fore grown only as an ornamental plant. Eventually, the peasant classes discovered its edibility when more desirable food was unavailable, but the development of a cuisine took several hundred years; the northern Europeans were reluctant converts, and it was not until the 1750s that tomatoes became widely used in Britain.

Despite its origins in subtropical Peru, the tomato is not, as one might think, a tropical plant. This is explained by the fact that it grew indigenously at some considerable altitude in the foothills of the Andes, and this also accounts for most of the plant's present-day environmental preferences.

Tomatoes like warm weather. For seed germination, a soil temperature of around 20°C is said to be ideal. For growing plants, fruit is abundant and flavoursome when night-time air temperatures are between 12°C and 20°C. If the night temperature falls below 12°C, stems and leaves are curbed at the expense of growing roots, while at night temperatures above 20°C, fruit will not appear—hence the difficulty of growing tomatoes near sea level in the tropics. The plant also has marked *thermoperiodic* requirements, in that for a bountiful yield, night temperatures must be a good 6°C lower than the daytime values.

The tomato plant grows well outdoors in many parts of Ireland and Britain except in cold, wet summers and, being 'day-neutral', it flowers successfully with either very short or very long days. This makes it possible to grow tomatoes in a frost-free environment during the short days of winter, as well as during the long days of the summers of more northern regions.

THE FIRST AND LAST OF THE WINTER FROSTS

7 November 2006 ⌒

The recent succession of fine days has been accompanied by cold, frosty conditions after dark, with the night air imparting a chill that has been foreign to us now for many months. Frost, however, is a very common phenomenon in most places around this time, although it would be a freak occurrence in July and August—one so unusual, in fact, that we could almost say it never happens. It is an interesting exercise in meteorology, therefore, to try to identify a date in autumn before which the first frost of the season seldom comes, and a corresponding date in springtime after which frost is unlikely to occur. One way of getting a measure on the situation is to examine the temperature records over a long period and calculate for various parts of the country the average date of the first frost of autumn and its last occurrence in the spring.

First we have to define what kind of frost we mean. In its simplest sense the term implies just any temperature of zero°C or below, but meteorologists are careful to distinguish between *air frost* and *ground frost*. There is often a marked change in temperature with height in the first few feet above the ground, and on a cool clear night the temperature at grass level may be several degrees lower than that at the height at which the air temperature is normally measured—four feet above ground level. Since ground frost is not too uncommon in *any* season of the year, it makes more sense when analysing the dates in the way described above to think in terms of the temperature of the free air, well clear of the grass and other vegetation.

By this criterion, we find that in the extreme southwest of the country, near the tips of the peninsulas of Cork and Kerry, the first air frost of the winter does not occur, on average, until early January. In the midlands of Ireland, the first air frost of the winter occurs in October, but areas adjacent to the sea are generally frost-free until December. The relative mildness of coastal regions is also in evidence in the spring. At the extreme tip of west Cork, the average date of the last 'winter' frost is 1 March. Elsewhere around the coastal strip it typically occurs in late March or early April, and in the midlands the last air frost of the season occurs, on average, during the first half of May.

Obviously all these dates vary from year to year, and are also affected by the lie of the local land. Altitude, for example, is an important factor; moreover, frost occurs more readily, and is therefore likely earlier in the autumn and later in spring, in sheltered valleys than in the more exposed parts of the surrounding countryside.

ROGUE WAVES AND STORMY SEAS

9 November 2006 ∾

There have been several instances on our western coasts in recent months of people being swept into the sea by a wave of clearly unexpected power and height. Such waves, although perceived in retrospect as 'freak', are in fact a common feature of chaotic seas.

Waves are creatures of the wind. Small wavelets, nurtured by a waxing wind, move in its general direction and increase in height as time goes by; the stronger the wind, the higher the resulting waves will ultimately be. But wave-height is also determined by what we call the *fetch*, the distance over which the wind has the opportunity to act. Given a steady wind, waves grow and grow in height as they travel over miles of ocean; but only after they have experienced sufficient fetch do they attain their full potential.

But waves have even more complexity. It often happens that a strong wind, having created a train of waves moving in a certain direction, begins to blow from a different quarter. The original train of waves survives as *swell*—waves no longer directly driven by the wind, but surviving by their own momentum as they decay over a rather lengthy period. The 'new' wind, on the other hand, creates its own new train of waves, capable of 'interfering' with the first.

Any number of individual wave trains may coexist in this way, the different components sometimes reinforcing one another to produce groups of very high waves, and at other times almost cancelling each other out to provide occasional zones of relatively quiet water. And every now and then, for just a minute or two, a large number of these individual components get exactly into step at the same place, and a wave of quite exceptional height enjoys a brief period of existence.

It was once believed that there was a pattern to all this activity, whereby individual waves were of gradually increasing height until a maximum was reached, and then the series all began again. One tradition had it that each tenth wave was the biggest, another that it was each seventh, and in *The Coming of Arthur*, Tennyson would have us believe it was the ninth:

   . . . And then the two
   Dropt to the cove, and watched the great sea fall,

Wave after wave, each mightier than the last;
Till last, a ninth one, gathering half the deep
And full of voices, slowly rose and plunged
Roaring, and all the wave was in a flame.

But regular patterns like this do not occur. Statistical calculations, however, broadly confirmed by observation, tell us that in a chaotic sea one wave in every twenty may be about twice the average height, one in every 1,000 three times the average height, and that one in every 300,000 waves may be a monster four times as big as any of the others.

MEASUREMENT BY LATERAL THINKING

10 *November* 2006 ～

Mr Gradgrind, you may recall, was the gentleman in Dickens's *Hard Times* who had strong views on education: 'Now, what I want is, Facts,' he instructed. 'Teach these boys and girls nothing but Facts. Facts alone are wanted in life. Plant nothing else, and root out everything else.'

Gradgrindism of this kind tends to produce young adults with an intellectual approach to life which might be described as unadventurous; a more enlightened environment, on the other hand, results in a facility for lateral thinking. Take, for example, the apocryphal student asked by his physics professor during an interview how he might determine the height of a tall building using a barometer.

Now the professor clearly had in mind the phenomenon whereby atmospheric pressure decreases gradually with height above the ground. At or near ground level this rate of decrease is about one millibar, or hectopascal, for every 30 feet or thereabouts. It follows, therefore, that if you have some means of measuring the difference in atmospheric pressure between two locations, you can estimate, with reasonable accuracy, their difference in altitude—in this particular case, a building's height.

The response, however, was quite unexpected. 'I would take the barometer to the top of the building,' replied the student, 'attach a long rope to it, lower the barometer to the street and then bring it up—measuring the length of the rope. The length of the rope is the height of the building.'

The nonplussed professor suggested his student try again; this time the approach was mathematical. 'I could take the barometer to the top of the building, drop it from the roof, and time its fall with a stopwatch. Then using the formula $S = \frac{1}{2} at^2$ (the distance fallen is equal to half the acceleration due to gravity multiplied by the square of the time elapsed) I could calculate the height of the building.'

Pressed for a less destructive methodology, the student's next suggestion involved the use of trigonometry: 'You could take the barometer out on a sunny day and measure the height of the barometer, the length of its shadow, and the length of the shadow of the building, and, using proportions, calculate the required height.' And he went on, without being asked, to provide another method. 'Alternatively,' he said, 'you could take the barometer and walk up the stairs; as you climb the stairs you could mark off the lengths, and this will give you the height of the building in "barometer units".'

But perhaps his last suggestion was his most ingenious: 'I would take the barometer to the basement and knock upon the janitor's door. When the janitor answered, I would say, "Dear Mr

Janitor, I have here a very fine barometer. If you will tell me just how high this building is, I will give you the barometer."'

GOD'S MYSTERIOUS WAYS

13 *November* 2006 ∿

Sometimes, while reading their Bibles, meteorologists try to imagine the means by which the Almighty might have engineered His multitude of miracles. A case in point is the famous parting of the Red Sea when the Israelites were *en route* to Canaan, hotly pursued by Pharaoh and his troops.

According to Exodus, 'the Lord caused the sea to go back by a strong east wind all that night, and made the sea dry land, and the waters were divided. And the children of Israel went into the midst of the sea upon the dry ground: and the waters were a wall unto them on their right hand, and on their left.' Then, when they were safely across, 'the waters returned, and covered the chariots, and the horsemen' of the poor Egyptians. Three theories have emerged from interdisciplinary research.

Perhaps, one theory suggests, the wind may have been responsible, more or less as suggested in the Bible. A strong wind blowing over a body of water for an extended period piles water up before it, rather like a ridge of soil before a bulldozer. One team of researchers has inferred from Exodus that the crossing took place at the northern edge of the Gulf of Suez, where the basin of the sea is narrow, long and shallow. Computers tell them that a 40mph wind blowing there for ten to twelve hours could have caused the shoreline to recede by about a mile, and the level of the sea to drop sufficiently to allow the Israelites safe passage. Then a sudden decrease in wind, or a change in its direction, might have allowed the waters to slosh back over the unfortunate Egyptians.

Another explanation is that the Egyptians may have been drowned by a tsunami associated with a nearby volcanic eruption underwater. Some experts point to the cataclysmic eruption of the Greek island of Santorini around 1600 BC, a timeframe, it might be argued, that fits the story recorded in the Bible.

And a third scientific explanation put forward is that the Israelites did not cross the Red Sea at all, but took a more northerly route across the delta of the Nile. Very high temperatures in this area often cause mirages to be seen, and images of the lagoons within the delta appear to float aloft, giving the illusion of a high wall of water all around. Perhaps the 'parting' through which Moses led his fugitives was just a watery illusion— although, in this case, what caused the Egyptians to be drowned?

But then most of these natural phenomena are so rare in that region that their fortuitous occurrence at the right time and in the right place to save the Israelites might be thought of as miraculous in itself; perhaps it is just that

God moves in a mysterious way
His wonders to perform.

A DISTINCTION WITH A DIFFERENCE

14 *November* 2006 ∿

Taxonomy is the art of classification—of painstakingly compiling what Keats called 'the dull catalogue of common things'. And meteorologists are experts at it. To the person in the street, for example, any careful distinction between rain and showers may seem very academic, since either or both may leave you wet enough to render such subtlety of terminology quite superfluous. But to weatherpeople, the distinction is important.

The difference has nothing to do with the length of time the precipitation lasts; the answer is in the clouds. Showers, by definition, fall from individual 'convective' clouds, like *cumulus* or *cumulonimbus*, while rain falls from a 'layer' cloud, like *altostratus*, which is spread more or less uniformly over a very large area.

The convective clouds that produce showers have a vertical structure, towering many thousands of feet into the atmosphere; frequently there is blue sky to be seen between them. Rain, on the other hand, forms in the relatively flat layers of *altostratus* associated with a weather front, which cover the sky like a thick blanket and make the landscape dark and threatening.

As it happens, shower clouds tend to be individual, local phenomena only a mile or two in diameter, so if the cloud is being carried along by a fresh breeze, the chances are that the shower associated with it will not last long in any particular place, but will have moved on somewhere else in a quarter of an hour or so. And since rain falls from an extensive area of cloud, the chances are it may take many hours to clear. But in calm conditions a shower cloud may sit over the one spot for many hours, and the result is still a shower—not rain. And likewise, rain may fall from one part of a layer of cloud and not from others, so the rain may come and go; but despite its intermittent character, it is still rain and not a series of showers.

The situation is further complicated when weatherpeople speak of 'rain showers'. They are not fudging a fine distinction, dearly held, but merely emphasising that the showers consist of drops of water, and not of sleet or snow or hail.

And finally, there is rain as we distinguish it from drizzle. This distinction has nothing to do with the *intensity* of the precipitation, and indeed a heavy drizzle may well be more wetting than light rain. The difference is based purely on dimensions; any water drop larger than a millimetre in diameter is, by definition, classified as rain.

Despite their fixation with taxonomy, however, meteorologists do not carry callipers to measure drop-size in their quest for accuracy. They use a simple rule of thumb: drizzle falls so gently that it will not disturb a water surface if it lands on it—but rain will cause a splash.

DEBATABLE LANDS ENGULFED BY SOLWAY MOSS

17 *November* 2006 ∾

'The devil damn thee black, thou cream-faced loon!' Macbeth's churlish rebuke to his messenger in 'the Scottish play' ranks truly as a classic of its *genre*—but then the news was bad. The hard-pressed Thane, you may recall, had been declared invincible 'Till Birnam Wood remove to Dunsinane', and now, it seemed, the wood was rapidly advancing on the castle.

Features of the Scottish landscape seem to make a habit of moving from one place to another, and of playing their part in the downfall of that country's kings. In the sixteenth century, for example, a fuzzy zone just north of Carlisle near Solway Firth was the subject of dispute between the Scottish and the English crowns. Here, between the rivers Sark and Esk on what were quaintly called 'the Debatable Lands', was a hill endowed with a multitude of springs, whose slopes were covered with an area of moss some two miles long and one mile wide. The springs maintained the moss in quagmire even in the driest summer—and formed the Solway Moss.

In November 1542, the armies of Henry VIII and James V, King of Scots, met on the Debatable Lands. The Scots were routed at a battle known as that of Solway Moss, since it is said that many of the retreating soldiers who escaped the English fell victims instead to the treacherous surface of the hillside bog. James, disheartened, died just two weeks later at Linlithgow, beginning the tragic saga of his daughter, Mary, Queen of Scots.

But two centuries later came another Solway Moss disaster in November. Heavy rains over several weeks had so engorged the Moss that 235 years ago today it breached the wall of solid earth that contained its eastern end; mud poured down into the valley, choked a stream, and caused a lake to form over an extensive area. A contemporary chronicler described it vividly:

Early on November 17th 1771, on a dark tempestuous night, the inhabitants of the plain were alarmed by a sudden and over-whelming irruption of the Solway Moss. Their dwellings were quickly surrounded with a thick black fluid, moving rapidly with large solid masses upon it, like floating islands, and deluging the extensive plain as it advanced.

Those who were aroused from their beds were obliged to fly, and left their cattle and their furniture a prey to the black, nauseous inundation. The dawn of the next morning exhibited to them an awful scene of horror and desolation; their houses and fields were completely buried in a stagnant lake, which continued to increase for several weeks, until it extended over about 500 acres, and was, in places, some thirty feet in depth. The twenty-eight families whose houses and farms lay ruined in this pitchy pool were reduced to great distress, having, as it was said, to begin totally their lives anew.

A FEAST OF FALLING STARS TO WISH UPON

18 *November* 2006 ～

Aristotle and his friends in ancient Greece were aware that a 'falling star' is not, in fact, a falling star at all. No matter how many of them 'fell' towards Earth, they noticed that the celestial population of heavenly bodies seemed to remain more or less the same; for want of a better name they called them 'things of the air'—or *meteors*. If you would like to wish upon one of these falling stars, and can tolerate a disruption to your normal sleeping pattern, tonight may be your opportunity.

Meteors are a kind of space pollution, tiny specks of inter-planetary dust that become visible by frictional incandescence as they encounter the rarefied air of the Earth's upper atmosphere. Occasional meteors may be seen on any clear moonless night, but at certain times of the year their number increases very noticeably, when Earth, in the course of its passage around the Sun, intersects a concentration of the meteoric dust. The result is what we call a 'meteor shower'.

Most of the tiny specks of dust that end up as meteors are scattered widely around the galaxy, but thick concentrations of such material are often associated with a comet. 'Showers' occur when the Earth, crossing the orbital path of a comet, passes through the very dustiest part of the comet's dusty trail. Every year from around 15–20 November, the Earth crosses the orbital path of Comet Temple-Tuttle, and the result is the 'Leonid' meteor shower, so called because the shooting stars appear to emanate from the eastern part of the sky in the vicinity of the constellation Leo.

Like others of its kind, Temple-Tuttle orbits the Sun on an eccentric, elongated path, and approaches our particular corner of the universe every thirty-three years. The really spectacular Leonid showers tend to occur near the time when Temple-Tuttle is at perihelion, at its nearest point to the Sun on its thirty-three-year odyssey. The Leonids were unmissable in 1799, 1833 and 1866, but disappointing on other such occasions—as, for example, in 1933 and 1999; indeed, like any meteor shower, although the Leonids occur annually on the same dates, their actual intensity in any particular year is difficult to anticipate with accuracy, making this an area where the astronomers, normally such faultless forecasters, have been caught out from time to time.

But techniques, apparently, have improved in recent years. We are told that the current Leonids are likely to exhibit a marked crescendo to about one or more every 30 seconds around 4.45am tomorrow, Sunday, morning. To spot them, you should look towards either the northeast or the southeast of the sky; observation will be facilitated by the fact that only the merest sliver of the waning Moon remains, but, like so many other things in life, it will all depend upon the weather.

THE ONION-LAYERS ABOVE OUR HEADS

21 *November* 2006 ∿

B enjamin Franklin was one of the first to investigate the structure of the upper atmosphere. In 1752, with his famous kite experiment in the middle of a thunderstorm, he demonstrated that lightning was electrical in nature. And for many years, this was virtually all that was known about what was happening far above our heads.

In due course, however, meteorologists became aware from frequent measurements on mountainsides that temperature decreases gradually with height. And until the beginning of the twentieth century it was generally assumed that this decrease went on indefinitely. But in 1902 a French meteorologist called Leon Teisserenc de Bort, using recording instruments carried aloft by hydrogen-filled balloons, found that above a certain height the temperature stops falling, and may even begin to rise again.

Meteorologists now think of the atmosphere in terms of four concentric spheres. They see it as onion-like, comprising four layers that alternate between warm and cold, each one being defined by its thermal characteristics. In ascending order from the ground, these four layers are the *troposphere*, the *stratosphere*, the *mesosphere* and the *thermosphere*.

The lowest layer of the atmosphere, adjacent to the Earth, is the troposphere. It is 10 to 12 kilometres deep at these latitudes, and contains nearly all the phenomena familiar to us as 'the weather'—the clouds, rain, thunderstorms and so on. Within the troposphere, the temperature decreases with height at a typical rate of about 5°C for every kilometre.

An abrupt change in the temperature regime takes place at what is called the *tropopause*, the boundary between the troposphere and the layer above, the stratosphere. Within the stratosphere temperature remains almost constant with height for a few kilometres, at around -50°C to -60°C, but then begins to increase again until, at 50 kilometres above the ground, it is not very different from that at ground level.

The next atmospheric layer in order of height, stretching from about 50 to 80 kilometres, is called the mesosphere, where the temperature falls off with height at about the same rate as in the troposphere. But the mesosphere is much deeper, so by the time we reach the *mesopause*—the top of the mesosphere—we are in a zone where the temperature is lower than anywhere else in our atmosphere, typically as low as -100°C.

Finally, above the mesopause, in the thermosphere, temperature again increases sharply with height, and 120 kilometres or so above the Earth it is warmer than at sea level. The warming at these high levels occurs because of absorption by the atmosphere of certain wavelengths of the Sun's radiation. But then the air is so very tenuous in these high regions that little energy is required to bring about a substantial rise, and indeed the whole concept of temperature as we experience it at ground level has very little relevance.

⏐ PLUMBING THE DEPTHS

22 November 2006 ⁓

W e who live on this planet are lucky to have our oceans and our seas. Many of our sister worlds are far too hot for liquid seas, and many worlds more distant from the Sun than we are eke out their existence in perennial ice. Earth is the only planet in the solar system to have oceans as we know them—vast bodies of liquid water exposed to the atmosphere above,

> *Calm or convulsed, in breeze, or gale, or storm,*
> *Icing the pole, or in the torrid clime*
> *Dark-heaving; boundless, endless, and sublime.*

As Byron clearly implies, even if he does not quote the figures, the topmost layers of the ocean vary significantly in their temperature, depending on the latitude. The coldest waters—those 'icing the pole' as he nicely puts it—register around -1°C; in the 'torrid clime', on the other hand, in parts of the tropics, the sea temperature may be 20°C or even 30°C. At the bottom of the ocean, however, the water temperature in the dark mysterious depths is remarkably uniform and low; regardless of latitude it is almost everywhere in the range 0.5°C to 1.25°C. This cold, bottom water has its origins in surface cooling near the poles, from where it sinks and spreads globally over the floor of the ocean to form a massive cold pool in Davy Jones's locker deep below the waves.

With relatively high sea-surface temperatures in the mid-latitudes and in the tropics, and a cold pool underneath, it is clear that there must be a decrease in water temperature with depth.

This decrease, however, is not gradual; the normal temperature distribution in the vertical consists of a layer of relatively warm water of near uniform temperature close to the surface, and then a very sharp decrease in temperature across a zone perhaps only one or two hundred feet in thickness, below which the water temperature is almost uniform again the rest of the way down. The shallow transitional layer is called the *thermocline*.

As you would expect, the contrast in temperature across the thermocline is greatest at the end of summer, after the surface waters have absorbed large amounts of energy from the Sun. But the distance below the surface of the thermocline itself also varies from place to place and from season to season. The thickness of the relatively warm surface layer depends to a large extent on storminess; strong winds mix the waters near the surface, and deliver a uniform temperature throughout the depth affected— so the stormier the weather, the deeper the warm surface layer will be. Typically at these latitudes, the surface layer might be 500 feet in depth after a stormy winter, while at the end of a calm summer, the thermocline might be a mere 100 or 200 feet below the surface.

DECEMBER'S DAYS ARE DARK AND DULL

1 *December* 2006 ∽

If there were to be an antithesis of the March paradigm, a month which came in as a lamb and made its exit like a roaring lion, it must surely be the November just deceased. With yesterday's storms fresh in our minds this morning, it may be hard to recall that the first days of that month were quite idyllic: calm, sunny, anticyclonic conditions gave us a perfect example of the All Hallows' summer famed in folklore. But what does December, the first month of the meteorological winter, have in store for us?

December days are dark and dull. There are fewer hours of sunshine than in any other month, although in one way this must be expected, since even if we were to enjoy cloud-free skies throughout, the long dark nights would curtail potential sunlight. In the event, the sky for most of the time is rather cloudy, and the combined effect is an average of only one hour of sunshine every day. John Greenleaf Whittier, of 'Barbara Frietchie' fame, summed up the picture nicely more than a century ago:

The sun that brief December day
Rose cheerless over hills of gray,
And, darkly circled, gave at noon
A sadder light than waning moon.

The typical December, even more than recent months, is often characterised by a persistent procession of active depressions making their way northeastwards close to the coasts of Donegal and Mayo. As a result, it is often a very windy month, with more than its fair share of stormy conditions. An average December brings gale force winds to our western coasts on eight or nine of its thirty-one days.

As winter gains in confidence and tightens up its grip, ground frost can be expected to be present on over half the mornings of the month in inland areas, but on only seven or eight days near the sea. The temperature on an average day rises to about 8°C in Ulster, and to about 10°C in the mild coastal regions of County Kerry. And speaking of the sea, those brave enough to contemplate a Christmas swim must be prepared for water temperatures of no more than 8°C or 9°C.

The highest rainfall occurs in the mountainous parts of Kerry where the December norm is about 400 millimetres, although parts of the midlands and east escape with less than 100 millimetres. And although January and February are the big months for snow, it comes along sometimes in December too; the records tell us that sleet or snow occurs on an average of three or four days this month in the northern parts of our island. Most of the time, however, these falls are not disruptive, and if they occur at all, they are most likely in December's closing days between Christmas and New Year.

THE MOON WITH ITS GAUZY VEIL

7 December 2006 ∾

Betwixt the frequent intervals of rain these evenings are patches of clear sky through which the Moon, now on the wane from full, is clearly visible. But sometimes, too, thin, fleecy, translucent clouds glide slowly past the ascending lunar disc, to give a fleeting glimpse of exactly what Percy Shelley had in mind:

And like a dying lady, lean and pale,
Who totters forth, wrapp'd in a gauzy veil,
Out of her chamber, led by the insane
And feeble wanderings of her fading brain,
The Moon arose up in the murky East,
A white and shapeless mass.

On such a night, variegated rings of light can often be seen in the cloud surrounding the softly radiant lunar face—rings that have a diameter only a few times that of the lunar disc itself. The display is a *corona*.

The brightest part, a whitish disc, is called the *aureole*. But if you look more carefully, you will see that nearest the Moon, the aureole has a bluish tinge, merging into yellowish-white, and that the whole disc has a brownish outer edge. And sometimes this whole ensemble is surrounded by larger and more beautifully coloured rings, in an outgoing sequence of blue-green-red that may be repeated more than once.

A corona is seen when cloud composed of water droplets partially obscures the illuminated Moon. The droplets cause *diffraction*, a process whereby the tiny waves of light are diverted from their original path by small obstacles in their way—in this case, the water drops. The effect of the intervening cloud is to divert in an observer's direction some of the rays of light which were originally headed a few degrees away from him or her—and which should, so to speak, have missed the observer altogether. The water droplets, in a sense, divert more moonlight in the observer's direction than he or she might normally be entitled to expect; this, happening over a large area, results in the bright circle of the corona.

The different colours occur because the process of diffraction affects some wavelengths of the composite, almost-white, light than it does others. The short blue wavelengths are bent very little; the longer red wavelengths are diffracted a good deal more. At the outer edge of the corona, the blue light is not bent sufficiently to reach our eyes, but the red light is—so light waves reaching us from that part of the cloud are predominantly red. Over most of the coronal disc, however, the different colours, first separated by the process of diffraction, are then superimposed again to blur into a uniform white.

The presence of a strong corona is an indication of great uniformity in the size of the water droplets making up a cloud. Moreover, the smaller the drops, the larger the coronal disc.

| THE WINDS IN PERSON

9 *December* 2006 ∿

The ancient Greeks personified their winds, imagining each as a named person endowed with the characteristics brought to mind by the breezes blowing from the various quarters of their world. Notos, for example, was the wind from the south, and was a sticky, slimy person; when portrayed in a painting or a sculpture he was provided with all the hallmarks of the excessive moisture he acquired on his passage to Greece across the Mediterranean.

The cold northerly wind was Boreas, a bearded old man. He was warmly clothed and portrayed holding a shell near his mouth, into which he blew to make his howling noise. Kaikias, the north-easterly, was portrayed holding an upturned shield half-full of hailstones, as if he were ready to rattle them down on the surrounding countryside. The other winds were also appropriately personified—like Apeliotes, the showery east wind, and Zephros, the gentle westerly.

The Native American Indians had a rather similar concept. They, too, personified the various wind directions, and their methodology in doing so is beautifully described by Longfellow in his epic poem *The Song of Hiawatha*.

The saga begins with Mudjekeewis, a brave warrior who slew Mishe-Mokwa, the Great Bear of the Mountains that had terrorised Indian tribes for generations. In gratitude, he was appointed father and keeper of the winds in perpetuity:

'Honour be to Mudjekeewis!'
Henceforth he shall be the West-Wind,

And hereafter and forever
Shall he hold supreme dominion
Over all the winds of heaven.

Mudjekeewis in turn divided the winds out among his family:

For himself he kept the West-Wind,
Gave the others to his children;
Unto Wabun gave the East-Wind,
Gave the South to Shawondasee,
And the North-wind, wild and cruel,
To the fierce Kabibonokka.

The three had characters ideally suited to their new responsibilities:

Young and beautiful was Wabun;
He it was who brought the morning,
He it was whose silver arrows
Chased the dark o'er hill and valley;
He it was whose cheeks were painted
With the brightest streaks of crimson.

Kabibonokka, on the other hand, had indeed a much tougher, crueller kind of character:

But the fierce Kabibonokka
Had his dwelling among icebergs,
In the everlasting snowdrifts …
He it was who sent the snowflakes,
Sifting, hissing through the forest,
Froze the ponds, the lakes, the rivers.

Shawondasee, however, was the nicest of the family, a languid, warm and easy-going soul:

Shawondasee, fat and lazy,
Had his dwelling far to southward,
In the drowsy, dreamy sunshine,
In the never-ending Summer.

It was Shawondasee who sent the migratory birds northwards in the springtime, and provided the right conditions for fertile crops and rich hunting. And it was he, too, who ...

Filled the air with dreamy softness,
Gave a twinkle to the water ...
Brought the tender Indian Summer
To the melancholy north-land ...

FOOD FOR CLIMATIC THOUGHT

11 *December* 2006 ∾

'Never was an event, stemming from factors so far back in the past, so inevitable yet so completely unforeseen.' Alexis de Tocqueville was writing of the French Revolution, but he might almost have had climate change in mind. The potential inevitability has been clearly outlined in the recently published *Stern Review on the Economics of Climate Change*. It warns that if we continue with business as usual, there will be sudden shifts in regional weather patterns threatening the livelihoods of tens of millions of people; there will be increased deaths from disease and malnutrition, and a possibility that 200 million people may become displaced by the rising sea level. There will be serious risks of violent conflicts in vulnerable areas of the world, and of major disruption to economic and social activity on a scale comparable to the two World Wars and the economic depressions of the early twentieth century.

A logical response to such a threat might be to address the basic causes of the problem, by reducing global greenhouse gas emissions by as much as possible. But the actual response has been summed up nicely by the respected environmentalist, James Lovelock: 'I am old enough,' he says, 'to notice a marked similarity between attitudes over 60 years ago towards the threat of war and those now towards the threat of global heating. Most of us think something unpleasant may soon happen, but we are as confused as we were in 1938 over what form it will take or what to do about it. Our response so far is just like that before the Second World War, an attempt to appease. The Kyoto agreement was uncannily like that of Munich, with politicians out to show that they do respond, but in reality playing for time.'

Since hope springs eternal, there have been discussions recently on a possible successor to Kyoto, to take effect when the present protocol expires in 2012. The outcome, de Tocqueville's 'unforeseen', is described succinctly in an editorial in *New Scientist*: 'The world's governments are sleepwalking to climate disaster. That much is clear from the UN's latest climate conference, held in Nairobi, Kenya. It is increasingly obvious that the minor diplomats and environment ministers in charge of climate negotiations have lost sight of the huge climatic forces that threaten future planetary security.'

And Dr Mike Hulme, Director of the Tyndall Centre for Climate Change Research at the University of East Anglia, brings the discussion back to de Tocqueville's inevitability: 'The idea of climate change reaches much deeper than simply the economic balance sheet. Differences in ideology, religion, psychology, governance and materialist aspirations all lie in wait to place obstacles in the way of any globally engineered pathway towards a serene climate. We will do what we can, and will try to do more, but climate change will not be "solved".

THE DOYEN OF THE DARWIN DYNASTY

12 *December* 2006 ∽

The Darwins were a clever lot. We all know about Charles, of course, whose theories about natural selection so revolutionised then-current notions about how evolution might take place. But three of Charles's sons achieved scientific eminence in their own right. Sir Francis Darwin, a botanist, became president of the prestigious British Association for the Advancement of Science. Leonard dabbled in astronomy and became president of the Geographical Society. And perhaps most famously, Sir George Darwin, as professor of astronomy at Cambridge University announced, plausibly at the time, that the Moon was a piece of the Earth which had broken loose many million years ago, leaving the Pacific Ocean as a deep scar of the separation.

Now, Charles's father, Robert, was an unassuming, albeit fashionable, country doctor in the town of Shrewsbury, but Robert's father, old Erasmus Darwin and grandfather to Charles, was perhaps the most versatile Darwin of them all. Erasmus was born 275 years ago today on 12 December 1731, and in addition to being an eminent physician and a poet of no mean reputation, he was one of the most distinguished scientists—or 'natural philosophers'—of his day.

His most important scientific work, *Zoonomia*, was published in 1796 and anticipated some of the ideas on evolution of his more illustrious grandson. Erasmus also proposed a viable system of plant and animal taxonomy—or classification—which gained a wide acceptance and might well have become the standard methodology had not that of Carl Linnaeus from Sweden become more widely known upon the Continent.

Erasmus made two significant contributions to the development of meteorology. In 1788, he published a scientific essay quaintly entitled *Frigorific Experiments on the Mechanical Expansion of Air*. In it he originated the idea that the lifting of air to great heights—by mountains, for example—was an important cooling mechanism which resulted in the condensation of moisture and hence contributed to the formation of clouds in the atmosphere.

At a more popular level, Erasmus had an interest in weather lore, and is widely credited with penning the familiar rhyme which lists virtually all the familiar signs of rain. It runs, in part:

Hark how the chairs and tables crack!
Old Betty's joints are on the rack;
Her corns and shooting pains torment her,
And to her bed untimely send her.
Puss on the hearth with velvet paws
Sits wiping o'er her whiskered jaws;

Loud quacks the duck, the peacocks cry,
The distant hills are seeming nigh . . .

Erasmus Darwin died in 1803, just six years before his famous grandson saw the light. Had the young Charles not brought such an exclusive notoriety to the family name, his mildly eccentric but multi-talented grandfather might well be better known today.

MEASURING A RAINSTORM WITH A TEACUP

16 *December* 2006 ∿

And every creek a banker ran,
And dams filled overtop;
'We'll all be rooned,' says Hanrahan,
'If this rain doesn't stop.'

How true for Hanrahan! It has been raining almost steadily now for over a month. The first half of November was relatively dry, but the second half made up for this in spades, bringing the total rainfall for November to around 150 per cent of normal values. And in December, with only half of it elapsed, the norm for the full month has already been exceeded in many places in the west of Ireland. The rain has been remarkable, not for any spectacular rates of fall experienced, but for its persistence and steadiness as each wet week followed on the heels of its damp and dreary predecessor, pouring yet more water on an already saturated landscape.

Most of us have an intuitive idea of the state of meteorological affairs from media reports and from our personal experience. The precise figures promulgated by Met Éireann, however, come from a network of several hundred rain gauges strategically situated here and there around the country. In former times, many of these were located at rural Garda stations, and readers familiar with Flann O'Brien's *The Third Policeman* will recall the frequent references to mysterious 'readings' carried out by the sergeant and his aide, MacCruiskeen; the latter records his figures carefully in his little black notebook, and on interrogation reveals them to the sergeant in terms like 'ten point six', 'seven point four on the beam' and 'a heavy fall at half past three'.

Gardaí now have better things to do. These days the readings taken in rural Garda stations are more likely to be those displayed by devices intended to discourage us from participating in any liquid form of Christmas cheer, or to detect those who may register 10 kilometres per hour above the monthly average on the local by-pass—so meteorologists have had to look elsewhere. But rain gauges are still maintained by a large and dedicated cohort of official, semi-official and voluntary observers, augmented by an increasing number of installations capable of registering the required data automatically.

Conceptually, the contents of a rain-gauge are, in Cicero's words, a *fluctus in simpulo*—'a storm in a teacup'. Although the gauge itself is scarcely bigger than a teacup, it must be assumed to have captured the whole character of any rainstorm extending over an area of many square kilometres. The statistics thus obtained may not be accurate to the nearest half a millimetre, but they do give a reasonable idea of the variation of rainfall from place to place, and an approximation to the real rainfall which is acceptable enough for most practical purposes. Meanwhile, on and off, the rain continues:

'We'll all be rooned,' says Hanrahan,
'Before the year is out.'

| BUMPS IN THE CLEAR AIR

18 *December* 2006 ∾

'Play it, Sam! You played it for her, now play it for me.' Who can forget those immortal lines from *Casablanca*, or the many times that Michael Caine has told us 'Not a lot of people know that'? And similarly enduring is Bette Davis's injunction in *All About Eve*: 'Fasten your seat-belts. It's going to be a bumpy night!'

Few seasoned air travellers, indeed, can have failed from time to time to 'hit an air-pocket', as it used to be described. The phrase, albeit rarely used nowadays, conjures up images of great celestial bubbles of invisible void floating in the atmosphere, into which, when least expecting to, the unwary aeronaut may fall. The reality, however, is rather less bizarre.

The phenomenon is properly known as Clear Air Turbulence, or CAT for short. It occurs mainly at high altitudes, and results from very sharp changes in wind strength or direction over short distances, either in the vertical or horizontal. The result can be a violent buffeting as spectacular as might be expected in association with the severest thunderstorm. CAT is all the more alarming in that it is often unexpected; unlike the turbulence of thunder-clouds, it is—as indeed its name implies—unaccompanied by any visual warning of its onset. Luckily, some 95 per cent of CAT reports are of the 'light' or 'moderate' variety, and only in a very small percentage of cases is the turbulence 'severe' or 'violent'.

CAT, unfortunately, is difficult to predict with confidence. It is also ephemeral in nature; it often happens that one aeroplane may hit a turbulent zone quite unexpectedly, while a following aircraft on exactly the same track may experience no turbulence at all. Forecasters reckon that there is always a background probability of somewhere around 2 per cent that CAT may be encountered, and their efforts at prediction concentrate on identifying those areas where the probability is likely to rise significantly above this figure and where any incident may be severe. They look most closely at the vicinity of the jet stream.

High above our heads most of the atmosphere drifts gently more or less from west to east. But here and there this eastward motion takes on a sudden burst of urgency, and the air is funnelled, as it were, into a very strong, concentrated tube of wind surging forward at 100 or 150 miles per hour. This jet meanders; sometimes it is relatively straight, but very often it undulates in a wavy U-shaped pattern, and it may lie anywhere between twenty-five and forty thousand feet above the ground. Wind speed drops off sharply at its edge, and also above it and below, providing a sharp contrast with the much quieter atmosphere around, and it is here, in these zones of very sharp transition, that CAT is most frequently encountered.

THE PROBLEM WITH ARTIFICIAL SNOW

19 *December* 2006 ∽

'*Mais où sont les neiges d'antan?*' asks François Villon in *Le Grand Testament*: 'But where are the snows of yesteryear?' This question has been asked frequently of late at many of the less elevated ski resorts in Alpine Europe. And personal observations on the dearth of snow have been confirmed by recent research results from a group of climatologists who have reconstructed past weather patterns over an area of Europe stretching from the Rhône Valley to Hungary; the region is warmer now for this season, they say, than at any time in the last 1,300 years.

None of this should surprise us very much, with so much talk of global warming. But stranger still are reports from these nether Alpine zones that it is too warm even to manufacture artificial snow. Surely artificial snow, being artificial, should not depend so much on temperature?

Snow of a kind can be manufactured in a snow cannon. The first requirement is water, in the form of water droplets in a very fine spray. The second is a plentiful supply of 'sublimation nuclei', tiny particles suspended in the air on which the embryonic snowflakes can be made to form. But the third essential, crucially in the present context, is a very low temperature; only when the air temperature inside the cannon can be reduced to -10°C, or even lower, can a plentiful crop of snowflakes be obtained.

Usually the most difficult part is finding the right nuclei on which to grow the snowflakes. Polystyrene granules have been tried, but they are less than satisfactory; the most effective nuclei, it seems, are tiny bacteria bred specially for the purpose. These are added to compressed air, which is stored at a low temperature to start with and then experiences another sharp drop in temperature with the reduction in pressure as it is suddenly released. The mixture of bacteria and cold air is combined with a very fine spray of water in the cannon, and as the mist crystallises, a shower of millions of bright fluffy snowflakes shoots forth onto the mountain slope.

Like most things artificial, man-made snow has disadvantages. It has a denser molecular structure than the real thing, so the snow has a tendency to compact into lumps of ice—so-called 'death cookies'—to trap the unwary skier. Secondly, its manufacture requires vast quantities of water, and the level of the local lake may be observed to drop alarmingly if the cannons are used over an extensive period. And thirdly, the process requires a great deal of energy; environmentalists point out that the energy used to produce artificial snow adds to the greenhouse effect, and by raising the global temperature still further it will create an even greater need for the artificial product. Satisfying this increased demand will add even more to global warming—and so on, as they say, *ad infinitum.*

| STORIES OF THE STAR

23 December 2006 ～

Just what are we to make of the story of the Christmas star? Cynics, of course, will tell you that it was all made up after the event to comply with the prophecy of the Old Testament: 'A star shall come forth out of Jacob, and a sceptre shall rise out of Israel.' Astronomers, on the other hand, point out in their meticulous way that every heavenly body they have ever come across was miles and miles away, and that it is nonsense to suggest that one could hover for days over some stable in the Middle East. But some scientists have tried to identify the Star of Bethlehem for what it may have been.

Unfortunately, it is not sufficient to calculate how the sky might have looked on 25 December in AD 1, or even in 1 BC. There is considerable uncertainty about the date of Christ's birth, and it is widely accepted that by our present calendar it could have been anywhere between 8 and 1 BC.

One's first thought might be that the star could have been a comet. It cannot have been Halley's Comet, since that is known to have appeared in 12 BC, and not again until AD 66, but a strong contender is Sui-hsing, the 'broomstar', so called because it had a tail that appeared to sweep the sky; it was recorded by Chinese astronomers as being visible for seventy days in 5 BC.

Others prefer the theory allegedly first proposed in 1603 by the astronomer Johannes Kepler. They note that every 139 years the Earth, Jupiter, Saturn and the Sun all lie more or less in a straight line, a phenomenon known as a 'triple conjunction' of the planets. Around the time when this happens, Jupiter and Saturn appear to trace out loops in the night sky against the background of the stars. A triple conjunction occurred in 7 BC, and on this occasion the planets were in the constellation Pisces, a coincidence which occurs only once every 900 years.

The significance of this unusual event, some argue, would not have been lost on the Magi who, eponymously wise, would have had knowledge of astronomy. They would also know that Pisces was a constellation associated in astrology with the Jewish people; that Jupiter was associated with royalty and brought good fortune; and that Saturn was the protector of the Israelites, who would herald a new age when the time was right.

Moreover, the biblical account in Matthew 2 implies that the Star was not an impressive spectacle, obvious to everyone. Perhaps it was not the single brilliant object that we think about, but rather an astrological signal in the sky, which would be of importance to those familiar with such things—but which would go unnoticed by the general population? But then, again, perhaps the suggestion of a miracle is just as plausible.

A CHRONOLOGICAL CONUNDRUM

30 *December* 2006 ～

Let me offer you a little conundrum this morning with which to tease your friends over the holiday weekend. Ask them which year was longer, 1751 or 1752, in a) England, b) Scotland and c) France. By way of a helpful hint, you might gently remind them of the provisions of Act 24 George II, steered through the British Parliament by Lord Chesterfield in 1751.

Lord Chesterfield's Act decreed that throughout Great Britain and Ireland the eleven days from 3 to 13 September 1752 should be omitted from the calendar. This had become necessary because the Julian Calendar, still then in use, was intrinsically inexact, and the dates had slipped out of phase with astronomical reality. Instructions for correcting the anomaly had been given nearly two centuries before, in 1582, in the Papal Bull *Inter gravissimas* issued by Pope Gregory XIII, and promulgated as they were by Rome, the new arrangements had been adopted almost immediately in nearly all Catholic countries throughout Europe. Protestant nations like Shakespeare's England, however, held it as almost a matter of principle to resist the proposal, no matter how sensible the changes it implied might seem. Only with the implementation of Act 24 George II did the British equinoxes, solstices and seasons revert neatly to their proper place.

But Lord Chesterfield's Act contained another overdue reform: that of fixing New Year's Day on the first day of January. Before the Norman Conquest, the year in these parts began on Christmas Day. Then in 1066 the Normans brought with them their custom of beginning the year in January, but this Continental affectation lasted for barely a century; from 1155 onwards New Year's Day was Lady Day, 25 March, and so it remained for most fiscal and legal purposes until 1751—except in Scotland, where 1 January had been adopted as the first day of the year in 1600.

Lord Chesterfield's Act, therefore, contained another nod towards Europe. It decreed, *inter alia*, that the day after 31 December 1751 should be 1 January 1752, thereby bringing England and Ireland into step with the Continent and establishing the convention which remains in place today.

These two provisions of Act 24 George II provide the answers to our question. In England and Ireland 1752 was longer than 1751; the latter ran only from 25 March to 31 December, and its three-month truncation well outweighed the loss of 1752's eleven days in September. In Scotland, 1751 had its full complement of 365 days but 1752 was deprived of those same eleven days. And in France, where the calendar was left alone because no reform was necessary, both years ran their full course; 1751 had 365 days, but 1752, being a leap year, lasted one day longer.

So the full answer to the question posed is a) 1752, b) 1751 and c) 1752.

AN ILLUSIONARY LATER DAWN

2 January 2007 ∾

The euphoria of the Christmas season has, as it were,

> *. . . melted into air, into thin air;*
> *And, like the baseless fabric of this vision,*
> *The cloud-capp'd towers, the gorgeous palaces,*
> *The solemn temples, are dissolved—*
> *Have like an insubstantial pageant, faded*
> *And left not a rack behind.*

Now, in the cold, dark, muddy days of early January, some of us may find our seasonal disillusion augmented by the fact that the stretch in the evenings, which has just begun to be apparent, is not reflected in a corresponding brightening of the early mornings—by any noticeable advancement of the time of sunrise. This asymmetry has more to do with clocks than with astronomy.

As the year advances towards the winter solstice around 21 December, the tilt of the Earth's axis in relation to its orbit results in a slowly diminishing daily dose of sunlight. Then from the solstice onwards, the number of hours of daylight gradually increases until midsummer. If we were to set our clocks exactly according to the Sun, to what is called 'Local Apparent Time' or LAT and by which noon each day is the exact instant when the Sun is due south, then the annual shortening and lengthening of the days would be symmetrical about the winter solstice. The latest sunrise and the earliest sunset would coincide on 21 December or thereabouts.

But in real life we do not set our clocks to LAT. Measured accurately by the Sun, the days turn out for various reasons to differ slightly in their length, being at certain times of the year a little longer, and at others a little shorter, than the precise interval we know as twenty-four hours. To avoid practical inconvenience, we 'pretend', as it were, that the days are all exactly twenty-four hours long; we use what is called *mean time*.

Because of this, our clocks are usually a little out of step with the Sun, apart altogether from the fact that we standardise our time by time zones. The difference between 'Sun time' and 'clock time' is called the 'equation of time', and varies rhythmically throughout the year; during December and January the effect of the equation of time is slowly to shift clock time a little forward each day as compared to real Sun time.

Try to imagine the effect of this upon the time of dawn. Once the winter solstice has passed, we *ought* to see earlier sunrises, but this trend is counteracted by the fact that our clocks are out of sync with nature; they show a progressively later time each late December morning than they ought to, which provides a trend for an *apparently* later dawn. Only near the end of January does the seasonal effect accelerate sufficiently to overcome this chronometrical illusion, and the mornings begin to become noticeably brighter.

A WORD IN YOUR SHELL-LIKE . . .

3 January 2007 ∿

'Will no one rid me of this turbulent priest?' raged Henry II a little after Christmas in 1170. And thus was set in train the sequence of events which led to the slaughter in his own cathedral of Archbishop Thomas à Becket.

One might imagine a similar outburst on the part of God in recent times when the activities of Professor Richard Dawkins are drawn to His attention. And as Dawkins shows in his recent book, *The God Delusion*, intervention of this kind would have been quite in character for God as portrayed a millennium or two ago in the Old Testament. But God, of late, has mellowed. He appears to have made no attempt whatever to rid Himself of Dawkins, despite the latter's sustained campaign to rid the world of God.

But there is another side to Richard Dawkins. Besides debunking the Almighty, he has produced several books that are rich in scientific odds and ends, among them *Unweaving the Rainbow*, in which he reveals that insects' ears are not barometers, but anemometers.

The function of an ear is to detect the tiny oscillations of the air we know as sound. These can be looked at as small variations in the barometric pressure, as the air is alternately compressed, and then released, by whatever makes the sound, the ripples of disturbance then propagating outwards from the source in all directions. The needle of a tiny, very sensitive barometer would swing up and down in phase with sound waves passing by, and this is exactly how a human ear behaves. The eardrum moves in and out with tiny pressure changes, its movement is relayed by nerve connections to the brain, and the brain decodes the message.

But insect ears are anemometers. Instead of tiny changes in the atmospheric pressure, sound waves can be thought of as little winds, the air oscillating to and fro in response to those changes in the pressure; the air moves in one direction as the pressure rises, and then returns again shortly after the pressure falls. While *our* 'barometer' ears have a membrane stretched over a confined space, insect 'anemometer' ears have either a hair or a membrane in a chamber open to the atmosphere; in either case it is blown back and forth by the rhythmic movement of the air.

A side-effect of this arrangement for the insect is that it is very easy for it to detect the direction from which any sound may come. We with our 'barometer' ears, on the other hand, have to calculate the direction of a sound by comparing the reports from *two ears*; the brain assesses the relative loudness of the sounds, and in a separate exercise compares the relative times of arrival of the sounds at each individual ear, and then makes an estimate of where the sound is coming from.

ROSIE REVISITED

4 January 2007 ∽

L aurie Lee's *Cider with Rosie* is well worth re-reading. An atavistic rural peace descends upon you as you immerse yourself in 'a world of silence, of hard work and necessary patience, of white roads rutted by hooves and cart-wheels, innocent of oil and petrol'. And of course you become re-acquainted with Rosie herself, the village girl who 'baptised' the adolescent author with her cidrous kisses beneath a hay-wain on a summer afternoon.

And there is meteorology galore. 'In the long hot summer of 1921 a serious drought hit the country. Springs dried up, and the usually sweet water from our scullery pump turned brown. For weeks the sky hung hot and blue, trees shrivelled, crops burned in the fields, and the old folk said the Sun had slipped in its course and that we should all of us soon die.'

There were prayers for rain, of course, but as the author says, 'the drought continued, prayer was abandoned and more devilish steps adopted.' The villagers had recourse to a persistent myth throughout the ages—and of course it is a myth—that gunfire prompts the clouds to pour forth rain.

Back in Roman times, Plutarch expressed the view that 'extraordinary rains generally fall after great battles'. As technology advanced, this perceived peak in rainfall came to be associated with gunfire, explosives and artillery. Napoleon is said to have believed that cannon-fire caused rain, being persuaded that the noise jostled minute cloud particles together, allowing them to coalesce and fall to Earth.

The *Irish Times*, too, had heard of this phenomenon. On 15 January 1915, its leader writer noted that 'two of the dominating factors in the lives of many of us, at present, are the rain and the war. It is only natural to try to trace them to a common source. It would appear not a little ridiculous to accuse Germany of being responsible for the present abominable weather in Dublin, but we believe that the charge can be made with some show of reason.'

The paper went on to observe that 'unusual amounts of rain have fallen both in Flanders and in Poland, where the firing has been heaviest. It may also be pointed out that the south of England, which is nearer to the actual field of battle than Ireland, has suffered from worse floods than any we have experienced.'

And the notion was clearly prevalent where Laurie Lee grew up. 'Finally soldiers with rifles marched to the tops of the hills and began shooting at passing clouds. When I heard their dry volleys, breaking like sticks in the stillness, I knew our long armistice was over. And sure enough—whether from prayers or the shooting, or by a simple return of nature—the drought broke soon after and it began to rain as it had never rained before.'

THE ART OF THE RAINMAKERS

5 January 2007 〜

Forecasting the weather is all very well, but the real trick is to control it, to turn the rain on when you want it and off when you do not. There have been many attempts to do so down the years.

In the US in the early 1890s, for example, Robert St George Dyrenforth exploited the persistent, mythical notion that gunfire and explosives enhance a tendency for rain. His methodology was to send explosives aloft in a balloon to be detonated by a time-fuse, thereby creating 'something in the nature of a vortex, a momentary cavern into which the condensed moisture is drawn from afar, after which the explosion may squeeze the water out of the air like a sponge'. He was sufficiently persuasive in promoting this extraordinary theory for Congress to vote him $2,000 to carry out experiments.

By coincidence, Dyrenforth's first trial in August 1891 produced a deluge. Others in succeeding weeks appeared to meet with moderate success, and Dyrenforth, for a while, was a nation-wide celebrity. But his subsequent attempts produced no rain at all, and he was pilloried and satirised as 'Dryhenceforth'. In due course he resumed his former practice of the law, a duller but safer occupation than his sally into hands-on meteorology.

Dyrenforth was followed some years later by Charles Malory Hatfield, who had similar ambitions but a different methodology —and rather more success. Hatfield turned professional rain-maker in 1905, and his methods were both enigmatic and impressive: a number of large towers 20 or 30 feet high would be erected where the rain was needed, each surmounted by a vat of boiling liquid to which Hatfield added his own secret chemicals—twenty-three of them in all.

A number of spectacular successes at 'tickling the clouds to tears', as some newspapers liked to put it, ensured that Hatfield's expertise was widely publicised. Over the next twenty years, one success for every dozen failures was more than enough to fuel a financially rewarding business. In 1924, the *Los Angeles Times* summed up Hatfield's public image rather well: 'Some think that Hatfield is merely a great showman; others think something less complimentary; but some, and always enough for his purposes, think him a man ahead of his time who can achieve what modern science and the US Weather Bureau say is quite impossible.'

No one ever convincingly showed that Hatfield was a fraud. In fact he seems to have been an honest, decent man who genuinely, if mistakenly, believed that his methods were effective, and who also had a unique talent for flamboyantly promoting his own perceived achievements. He continued in business until the early 1930s, and then retired to live quietly until his death in 1958—a controversial and elusive figure, but one of the most colourful ever to have featured in the chequered history of pluviculture down the years.

MAKING RAIN THE MODERN WAY

6 January 2007 ~

C onsiderable effort has been expended over the years on techniques to enhance natural rainfall in areas susceptible to drought. It is a daunting task; the energy involved in even very small-scale atmospheric processes is so vast that any human input can only act as a catalyst, with a view to triggering something which was, perhaps, about to happen anyway.

A case in point is the technique of 'cloud-seeding'. It originated in 1946 when an American chemist called Vincent Schaefer discovered that if powdered carbon dioxide, or 'dry ice', were dropped into a bank of cloud from an aeroplane, it sometimes resulted in a fall of rain. The reason is that the particles of dry ice, or the silver iodide that is sometimes used nowadays, act as a catalyst to encourage water droplets in the cloud to change into natural ice crystals, which in turn enhances the cloud's ability to rain. In the years that followed, the 'crop-dusters'—those adventurous aviators who earned their living by spraying crops from the air on the vast plains of the United States—found a new outlet for their skills; they were asked by the ever-thirsty farmers and ranchers to concentrate on a new target, spraying the summer cumulus clouds and thereby precipitating a spate of popular magazine articles in which their glamorous activities were vividly described.

In the early days of cloud-seeding, techniques suffered from the difficulty of deploying the seeded particles throughout a sufficiently large volume of a cloud to render them effective. More recently, cloud-seeders have tried to allow the clouds themselves to do the work—by spraying the particles from an aircraft flying *underneath* potential shower clouds, and letting the updrafts in the clouds effect dispersed delivery.

A difficulty arises, however, when trying to assess the efficacy of such a process: rain may well occur after the operation—but who is to say that it might not have happened anyway, even without this human intervention? During the early 1980s, detailed studies were carried out in Australia, Israel and the US, to test the effectiveness of cloud-seeding, and the results were analysed by the British cloud physicist Sir John Mason. He concluded that 'In the case of the Tasmania and Florida experiments the evidence does not provide strong support for a positive seeding effect. Statistical evaluation of the Israeli experiment, however, provides much more convincing evidence, and suggests an average increase in rainfall due to seeding of about 15 per cent. But why the clouds in Israel should be more responsive to seeding than rather similar clouds in other parts of the world is not immediately clear.'

Neither is it clear how Irish clouds might react to such an onslaught. Luckily it is only very rarely that such an experiment might seem in any way attractive.

AN ACOUSTIC ANOMALY EXPLAINED

8 *January* 2007 ∽

Most sounds, as we know, become inaudible at a relatively short distance from their place of origin, as the sound waves *attenuate*, or weaken, and ultimately die away.

But centuries ago it had been noticed that the sound of gunfire was often heard, not only in the immediate vicinity, but also in an outer ring 60 miles or more from the source, with a zone of silence in between.

Samuel Pepys in his diary, for example, makes reference to this phenomenon. In early June 1666, the Anglo-Dutch Wars were in full swing and a naval skirmish, the so-called Four Bay Battle, took place along the continental coast between Dunkirk and Ostend. On 2 June, Pepys wrote: 'I went on shore with Captain Erwin at Greenwich, and into the parke, and there we could hear the guns from the Fleete most plainly.'

That evening Pepys heard about passengers aboard the yacht *Katharine* in the English Channel, 'who saw the Dutch fleete on Thursday and ran from them; but from that hour to this hath not heard one gun nor news of any fight'. And the Governor of Dover Castle dismissed reports of gunfire heard in London as 'only a mistake for thunder'.

On 4 June, Pepys summed up the situation as he saw it: 'It is a miraculous thing that we all Friday, Saturday and yesterday, did hear the guns most plainly, and yet at Deal and Dover they did not hear one word of a fight, nor think they heard one gun. This, added to what I heard about the *Katharine*, makes room for a great dispute in philosophy—how we should hear it and they not, the same wind that brought it to us being the same that should bring it to them: but so it is.'

Not until the early years of the twentieth century was this anomaly explained. Listeners in the inner zone hear sound waves which have travelled directly towards them through the lower atmosphere, but the existence of the outer zone implies that waves originally moving *upwards* from the source are subsequently redirected downwards to reach the Earth a great distance away. And the only reason why sound waves should so dramatically change their direction of propagation would be the existence of a region of relatively high temperature high above the Earth.

During the 1920s meteorologists took accurate measurements of the dimensions of the 'silent' and 'noisy' zones associated with prearranged loud explosions, and also measured the length of time taken by the sound to reach a selection of different sites. From these data, they calculated that a warm layer must exist in the atmosphere between 20 and 40 miles above us, with temperatures not too different to those we experience here near the Earth's surface. These conjectures have been amply confirmed in the intervening years.

THE STRANGE BOOMS OF THE BARISAL GUNS

9 January 2007 ~

In certain places around the world, mysterious booming sounds are heard from time to time near the sea or other areas of open water. They first came to general notice towards the end of the nineteenth century near the town of Barisal on the Ganges Delta, and since then, wherever they might occur, they are known by the generic term of 'Barisal guns'. Their most intriguing characteristic is that they have no obvious explanation.

One of the earliest first-hand accounts of the original Barisal guns appeared in the scientific journal *Nature* in January 1896: 'I first heard the Barisal guns in clear weather on my way to Assam from Calcutta through the Sunderbans. They came sometimes singly, at other times two or three or more in quick succession. I was told the reports were heard all over the Sunderbans, and that several experts had failed to account for them.'

But sounds of a somewhat similar nature were reported even earlier here in Ireland. *Smith's Natural & Civil History of Kerry*, written in 1756, for example, referring to the coastal townland of Minegahane, tells us that 'the most remarkable curiosity of the place is the prodigious noise made at certain seasons by the sea, somewhat like the firing of a cannon, which may be heard at great distance; this generally precedes a change of the wind and weather and frequently happens towards the approach of a storm.'

Another Smith, Rev. W.S. Smith, described a similar phenomenon in Ulster in the 1890s:

> For many years after my settlement here from England, I heard at intervals when near Lough Neagh, cannon-like sounds; but not being acquainted with the geography of the distant shores, and the possible employments carried on, I passively concluded that the reports proceeded from quarrying operations, or on fine summer days from festive gatherings in Derry or Tyrone. In time I came to understand, however, that it was not from the opposite shores but from the lake itself that the sounds proceeded. I have heard the reports on probably twenty occasions during the present year. After questioning many of the local residents I extended my enquires to the fishermen, but they could assign no cause, and the origin of the water guns remains a mystery.

One eminent scientist, after years of studying reports of booms like these around the world, concluded that such sounds 'cannot be due to a stormy sea, because they are most frequent when the sea is calm; if their origin were atmospheric, they would not be confined to specific regions; nor can they be connected with artificial noises for they are heard by night as well as by day and in countries where the use of explosives is unknown. There remains the hypothesis that they are of seismic origin'. Even today, as far as I know, the Barisal guns are an enigma.

A SAINT REMEMBERED FOR A STORM

15 January 2007

Storms at least comparable to those troubling our shores at present were common in the middle of the fourteenth century. Indeed it was a troubled time all told. The winter of 1348–49 brought the bubonic plague, the infamous 'Black Death', which wiped out half the population in parts of continental Europe; it ravaged much of the eastern half of Ireland equally effectively. The following years were remarkable for super-abundant rainfall, and also for a succession of fierce gales culminating in the arrival of 'St Maury's Wind' in 1362.

No one has suggested that Maury himself, or more conventionally St Maurus, was in any way to blame for the catastrophe. This good and holy man was born about AD 510, son of a Roman nobleman, and when he was about twelve years old his father placed him in the care of Benedict, founder of the Order that still bears the latter's name. Within that fold, apparently, Maurus advanced in piety and learning, and became a model of perfection to his brethren, most especially when it came to the virtue of obedience. He also learned some quite extraordinary skills.

When St Placid, for example, another inmate of the monastery at Subiaco, went one day to draw some water from the lake, he carelessly fell in. Maurus, having the advantage of a vision of the happening in his cell, hastened to the lake and, scarcely pausing at the edge, walked upon the waters as if on dry land. He seized the, no doubt by now, decidedly 'unplacid' Placid by the hair, and dragged him to the shore.

Maurus ended up in France, where he founded the famous abbey of Glanfeuil. He ruled over it as abbot for nearly forty years, before expiring, with full ecclesiastical honours, on 15 January 584. The fifteenth of January, therefore, became St Maury's feastday, and hence the eponym by which we recall the famous storm.

St Maury's Wind caused widespread devastation through the countryside, and particularly in Dublin City. It has been immortalised in an epic poem by one John Harding—a poem, it must be said, which must be appreciated more for its putative historical insight than for its literary merit or its author's rhyming skill:

In that same year—'twas on Saint Maury's Day—
A great wind did greatly gan the people all affray;
So dreadful was it then, and perilous,
And 'specially was the wind so boisterous,
That stone walls, steeples, houses, barns and trees
Were all blown down in diverse far countrees.

Another contemporary chronicler described St Maury's Wind a little more succinctly as 'a vehement wind, which shook and threw to the ground steeples, chimneys and other high buildings, trees beyond number, divers belfries and the bell tower of the Friars Preachers in Dublin'. By either version, it was quite a storm.

THE CITY COUNCIL'S WEATHERMAN

18 *January* 2007 ∿

Oyez! Oyez!

The advice of an old and seasoned meteorologist to a rookie colleague on the delivery of weather forecasts on the radio or television might well resemble that of the pernickety Hamlet to the players about to take the stage in Elsinore: 'Speak the speech, I pray you [. . .] trippingly on the tongue; but if you mouth it, as many of your players do, I had as lief the town-crier spoke my lines.'

Believe it or not, in Dublin in the late sixteenth century it really was the Bellman, an equivalent of the town-crier, who gave notice of impending gales. At a meeting of the Dublin City Council held in 1577, it was decreed 'that Mr Mayor and his Brethren shall devise and appoint whereby a bell shall be tolled in time of great tempest and storms, so that every well-disposed citizen may be remembered to pray for his neighbours who may be in danger upon the seas; and that Mr Mayor and his aldermen shall decide what allowance or reward shall be given to him that shall take pain in the knolling of the said bell.'

Twelve months later, one Barnaby Rathe was appointed Bellman to the city to undertake this task, and he continued in office until near the turn of the century. Unlike today's promulgators of warnings of windy weather, however, Barnaby was also 'master of the beggars', and as such was required to 'do his duty in killing such swine as he shall find in the streets, and ridding the city of its vagabonds'.

But to return to Hamlet and the forecasters, the prince's further advice on good delivery, if you remember, was that 'in the very torrent, tempest, and—as I may say—whirlwind of passion, you must acquire and beget a temperance that may give it smoothness'. Weather forecasts are masters of such temperance; they strive incessantly to be precise, concise and understandable.

To meteorologists, for example, a 'gale' is not just a strong wind; it is a wind whose average speed, measured over ten minutes and ignoring gusts, exceeds 39mph, or 62km/h. And the term 'gust' itself has a specific meaning: it is a sudden but very short-lived rise in wind speed, a transient phenomenon not to be confused with 'squall', which denotes a more prolonged increase in wind strength which may last for several minutes.

The word 'storm', on the other hand, is used in two different ways. In one sense a storm is a deep depression of particular severity, like those we have experienced of late. But the term 'storm force' is used to describe winds of a certain strength; they are those of Force 10 or 11 on the Beaufort Scale, blowing at speeds of between 55mph and 72mph, or 88km/h to 115km/h.

THAT'S HIS STORY

22 *January* 2007 ∽

Today is the centenary of the birth of Douglas Corrigan. He was not born great, but he achieved iconic status in his lifetime, and opinion is still divided as to whether this greatness was achieved, or simply thrust upon him by a navigational mistake. His place in aviation history is that of the variously incompetent, or devious, 'Wrong-Way' Corrigan.

Born in Galveston, Texas, on 22 January 1907, Corrigan qualified as a pilot and worked on the preparation of Lindbergh's *Spirit of St Louis*. He subsequently made no secret of his wish to emulate the latter's great achievement, but seemed resigned to the intransigence of the American authorities, who opposed non-stop solo flights across the ocean on the understandable grounds that the risks involved were unacceptable. Then, on 8 July 1938, Corrigan departed Long Beach, California, for New York on what he claimed would be a round-trip flight. He landed at Roosevelt Field on 9 July, and departed on the seventeenth for the return journey to Los Angeles.

According to himself, once in the air he misread his compass by observing the wrong end of the needle and therefore headed east instead of west. He blamed the weather for compounding his mistake; for almost half the journey, he said, thick cloud above and below obscured his view, so that he had no way of taking bearings from landmarks or the Sun. By the time he discovered his 'error', it was too late to turn around and he landed at Baldonnel Aerodrome after a flight of 28 hours 13 minutes.

Corrigan's version of events is supported by the fact that he neither sought nor was given any weather information for the North Atlantic, but received a full briefing on the route to California. Moreover, the only map aboard related to the west-ward route, and he carried no passport or related documents. But the aviation authorities remained convinced they had been duped, and that Corrigan really intended to fly the Atlantic all along.

The now famous 'Wrong-Way' Corrigan was duly fêted in Dublin and sailed from Cobh a few weeks later to return home by more conventional means to the traditional ticker-tape welcome in New York in early August. He did well out of his adventure, writing a best-selling autobiography, *That's My Story*, and star-ring as himself in a film called, anomalously, *The Flying Irishman*. In due course he bought an orange farm in California and lived there quietly for the remainder of his life.

But right up until his death in December 1995, Corrigan's answer to the perennial question was: 'The only story I know is the one I've been telling all the time. I've told it so many times, I'm beginning to believe it myself. Why should I change it now?' And to this day, 'Wrong-Way Corrigan' is a stock colloquial phrase in popular culture.

METEOROLOGY'S SOURCE OF ECCENTRICITY

25 January 2007 ∿

'In *that* direction,' says the Cheshire Cat with a wave of a paw to Alice as she tries to come to terms with Wonderland, 'lives a Hatter; and in *that* direction … lives a March Hare. Visit either you like; they're both mad.'

The key to the hare's legendary eccentricity in March lies in its courtship rituals. As in the case of some humans, the amorous enthusiasm of competing males may result in strange antics, like bounding to and fro, wild kicking, and the curious spectacle of protagonists standing on hind legs and appearing to box with one another. This unrestrained exuberance during the rutting season, compared to other times of year, has led to the expression 'mad as a March hare'.

Mad Hatters, on the other hand, acquired their reputation from quite a different source. In days gone by, mercury was often used to cure the felt for hats. But mercury, for all its silky sheen and the strange attractiveness of its paradoxical liquidity, is not a pleasant substance; it is highly toxic, and over time the fumes inhaled by hatters could cause neurological disorders, resulting in confused speech, distorted vision and in more severe cases, psychotic symptoms like hallucinations. By the indelicate terminology of the day, affected persons were described as mad.

Over the centuries, however, mercury has served meteorologists well as the working fluid in their thermometers. When physicists started to experiment with different ways of measuring temperature, it was obvious that the key to the problem was the fact that a change in temperature brings about a change in volume. In due course they hit upon the idea of a bulb filled with liquid, attached to a narrow tube of glass; the changing level of the fluid in the tube would reflect variations in the temperature.

The two most popular liquids used at first were alcohol and water, but both had disadvantages. Water solidifies at 0°C, so a thermometer which uses it as a medium will not function in sub-zero temperatures. Alcohol, on the other hand, remains liquid until its temperature is well below -100°C, but it boils at 78°C, another obvious limitation. Mercury, however, was ideal; not only does it have a conveniently large range of liquidity (from -40°C to 357°C), but it also reacts quickly and uniformly to any change of temperature. As an added bonus, the thin silver thread in the glass tube is easy to observe. But might the substance explain the eccentricity of some members of the meteorological profession?

In any event, the days of the mercury thermometer are numbered. The substance is now recognised to be a prime health and environmental hazard, and its use in meteorological instruments has been prohibited in Norway since 1998. Similar restrictions apply in France, Sweden, Denmark and the Netherlands, and a pan-European ban is likely in due course.

THE ELUSIVE PIMPERNEL
OF METEOROLOGY

27 January 2007 ∽

Every now and then in the learned scientific journals we read of a new theory to explain ball lightning. But then we hear no more, until we see yet another theory six months later. The phenomenon remains a great unsolved mystery of meteorology—a riddle, wrapped in a mystery, inside an enigma, encapsulated in a little glowing sphere.

Ball lightning may be white, red or orange in colour, and typically approximates a tennis-ball in size. The apparition usually occurs in thundery conditions when ordinary lightning has been seen as well, and may be accompanied by a hissing sound and a distinctive smell. It often drifts along some feet above the ground, as if wafted in a current of moving air, and then, after several seconds, it disappears as quickly as it came. It seems to be relatively harmless, occasionally causing a little damage but rarely of a serious nature; people coming into contact with ball lightning have described the sensation as like receiving a glancing blow, albeit somewhat painful.

In scientific circles, ball lightning was for a long time dismissed as more or less imaginary; some suggested, for example, that it might be an optical illusion, caused by an after-image on the retina immediately following a stroke of lightning. But after many thousands of more or less authenticated sightings over the years, most scientists now accept that the effect is real; but they are still nonplussed about its origins. It poses a litany of paradoxes: it glows like a 100-watt bulb, yet has no obvious source of power; it emits very little heat, yet it has been known to melt glass windows;

and it floats like a ball of gas, yet hangs together like a ball of liquid.

It has been suggested that a fireball may be some form of 'brush discharge' of static electricity; or that it may, perhaps, be a vortex of air containing a dense concentration of inexplicably luminous gases. A more convincing theory was proposed some years ago by a New Zealand scientist, Dr John Abrahamson, who recalled that silicon, a common constituent of many soils, has the unusual property of being unstable, and oxidising at high temperatures; it reacts with the oxygen in the atmosphere and releases chemical energy in the form of light and heat. His theory was that when a bolt of lightning strikes the ground, the intense heat vaporises silicon and a further chain of events provides the fireball's glow.

Until now, efforts to produce ball lightning in the laboratory have been fruitless. But recently, by passing a very high current through thin slivers of silicon, a team of scientists from the Federal University of Pernambuco in Brazil have succeeded in producing luminous orbs the size of ping-pong balls that have persisted for up to eight seconds. Dr Abrahamson, understandably, is said to be delighted.

SAVED BY THE RAIN

29 January 2007 ∿

D own through the years there have been many references in this column to individuals who have claimed to produce abundant rain in places afflicted by persistent drought. Their methods and their motivations have varied widely.

In the US in the 1890s, for example, George Dyrenforth used high explosives detonated by a time-fuse; Clement Wragge in Australia in the early 1900s placed his trust in batteries of artillery pointed at the clouds; and in the 1920s, Charles Hatfield prepared large vats containing secret chemicals whose fumes, he claimed, enticed the Californian clouds to part with water. But centuries before, St James the Ascetic used more traditional techniques, and for quite different reasons.

Yesterday, 28 January, was James's feastday. His career is described in detail in the *Apophthegmata Patrum*, the *Sayings of the Fathers*, an ancient collection of anecdotes about the more prominent holy men and hermits who frequented the Egyptian deserts in the fourth century AD. James, by all accounts, was a devout young man who at an early age removed himself from worldly things to settle in a cave near the village of Porphyrianos; here, we are told, 'he struggled with long fastings and with numerous prayers'.

There were those, however, who tried to divert James from his asceticism. Evil men sent a harlot to his cave who tried to tempt him, but the stalwart James preached to her, and reminded her of the fires of hell and of eternal punishment; the young lady duly repented of her ways and returned to the city, praising God.

James was less successful against a second onslaught. A certain man, apparently, had a daughter possessed by demons and he brought her to the saint for healing. James prayed and immediately freed her from possession, but the girl's father, fearful that the demons might return, left the girl and her younger brother in the saint's care in the cave. James, in due course, 'was defeated by desire'; he seduced the young woman, then killed her and her brother and threw the bodies into a nearby river. After this moment of madness, James was understandably afraid that his abhorrent sin might be discovered, and was also overwhelmed with great remorse. He shut himself in a tomb to weep and pray.

God eventually took pity on the entombed James. He had the land smitten by a devastating drought, and then informed the local bishop in a dream that only James had the power to solve the problem. The bishop with all the people went to the saint's tomb and begged him to pray for rain, and when he reluctantly agreed, the surrounding lands were deluged with abundant supplies of badly needed water. Moreover, it was clear to the joyful James that God must have pardoned him for his misdemeanours, and in this more cheerful frame of mind he continued with his rigorous regime until the day he died.

VARIATIONS ON THE GROUNDHOG THEME

1 *February* 2007 ✑

B e mindful on the morrow of Sir Thomas Browne. Sir Thomas was an English author and physician of great repute, *qui floruit*, as they used to say, in the reign of Charles ii. Although competent by all accounts in medicine, he is better remembered for his writings, 'a collection of opinions on a vast number of subjects more or less connected with religion, expressed with a wealth of fancy and with wide erudition'.

Sir Thomas did not shy away from meteorology. We are indebted to him for passing on a long-range forecast, which in its Latin form he quotes as follows:

Si sol splendescat clare, Maria Purificante,
Major erit glacies post festum—quam fuit ante.

The message, roughly speaking, is that if the Sun shines brightly on the feast of Candlemas, tomorrow, 2 February, there will be more of winter to follow than has gone before; in other words, we can expect a 'second winter'.

Sir Thomas's principle has many variants. A version in English, for example, runs:

If Candlemas Day be fair and bright
Winter will have another fight;
But if Candlemas brings clouds and rain
Winter is gone and won't come again.

The Germans, on the other hand, added a dash of Beatrix Potter to the mix. Their legend has it that the hibernating hedgehog, or sometimes the badger, creeps from his den at noon on Candlemas to see if he can find his shadow; if there is no shadow, he stays out, but if the Sun is shining the animal returns to his den and remains there for six weeks until the inevitably imminent spell of cold, wintry weather has passed away.

It may well have been German emigrants to the New World who brought Candlemas to Pennsylvania, but the tradition was transposed to the behaviour of a large, local rodent called the groundhog. The groundhog also hibernates, and on 2 February each year it allegedly emerges from its burrow to see if it will cast a shadow; if the Sun is shining, it slinks back into its den immediately, and stays there until the ice and snow have come and gone.

The citizens of Punxsutawney, Pennsylvania, have made their own of the prognostic skills of the groundhog, or 'Punxsutawney Phil' as they like to call him, and every 2 February, 'Groundhog Day' is marked by great festivities and celebration. The film of that name, popular in the cinemas some years ago, told the story of a TV weatherman who was sent to Punxsutawney to report on the event; the tale unfolds to the tune of the catchy number *I'm Your Weatherman*, and has our hero condemned to relive the same day over and over again *ad nauseam*, a sentence that undoubtedly had its disadvantages but should have worked wonders for his apparent skill at weather forecasting.

| THE SCEPTICS' RATIONALE

2 *February* 2007 ∿

No one now denies that planet Earth is getting warmer year by year, and it has become almost Luddite in scientific circles to deny that human activity is the cause. But those few who still have reservations base their scepticism on the demonstrable fact that the climate of the Earth changes over the years anyway, without any help from CO_2 or human beings. We are subject to the whims of a climatic pendulum, by which several times during every million years great glaciers move southwards across the continents in a colder world, only to retreat polewards again as the global temperature increases. On a shorter timescale, the benign Mediaeval Optimum in Europe a thousand years or so ago was followed by the Little Ice Age, from whose rigours we have only recently, in climatic terms, recovered.

Climatic change can be caused by variations in the energy output of the Sun. Until comparatively recently the *solar constant* was assumed to be just that, but it is not exactly so; the amount of energy available to fuel the world's weather changes over anything from decades to millennia.

But even with the Sun's output constant, the amount of its energy absorbed by Earth may vary for several reasons, the most important of which are related to idiosyncrasies in the orbital characteristics of the Earth. Our path around the Sun, for example, is not a perfect circle, but an ellipse. Moreover, the shape of this ellipse is not a constant; it goes from being rather elongated—when the astronomers say its *eccentricity* is high—to very near a perfect circle, and then extends itself again.

Another element of this climatic jigsaw puzzle is the *axial tilt.* We all know that the Earth's axis is tilted at an angle of 23½ degrees to the plane of the Earth's orbit around the Sun—but this figure is not constant either; it varies over thousands of years from about 25° to 22°, a change in what astronomers call the *obliquity of the ecliptic.* Since it is the axial tilt which gives us seasons, it follows that the amount of tilt affects the severity of winters; when axial tilt and orbital eccentricity combine to do their worst, we get an Ice Age.

The global climate is also affected by the amount of volcanic activity on the planet at any given time. More volcanoes mean more dust in the atmosphere, and this interferes with the transparency of the air to radiant energy from the Sun—and thus affects our weather.

Scientists have devoted much effort to assessing these processes, and others, to establish which are most significant. Ultimately, of course, the best way to test any scientific hypothesis is to initiate a possible cause and then observe the effect. Humanity, albeit unintentionally, is currently conducting precisely this experiment.

| THE IPCC'S SECRET WEAPON

3 *February* 2007 ∽

Calvin Coolidge, US President from 1923 to 1929, had a great dislike of ostentation. 'Men in public office,' he declared, 'should substitute for the limelight that light that comes from burning the midnight oil.' Consistent with this paradigm, he himself had a capacity for quiet, hard work, and a liking for trenchant phrases, often humorous but invariably terse. His taciturn character prompted Dorothy Parker, when told of his demise in 1933, unkindly to enquire: 'How could they tell?'

The same question could be asked about the Intergovernmental Panel on Climate Change, the IPCC, who yesterday issued their Fourth Assessment Report giving the direst warnings yet about the prospects for the global climate. How do they know what the climate of the world may be more than a hundred years from now?

The answer lies in CGCMs, or Coupled General Circulation Models. CGCMs, like those used to produce the daily weather forecast, are based on a mathematical model of the atmosphere—a description of the behaviour of the atmosphere in terms of mathematical equations. The computer starts with a description of the current atmosphere and, applying the equations at thousands of 'grid-points', works its way forward, step by step, to some time in the future.

But there are important differences between the *forecast* models, used to predict the weather just a day or two ahead, and the *climate* models, which predict the general state of the Earth's climate decades hence. Forecast models have to predict in great detail the behaviour of relatively small-scale atmospheric phenomena, and are therefore very sensitive to tiny errors in the accuracy of the data which define the state of the atmosphere at the beginning of the exercise.

Climate models, on the other hand, are concerned only with broad trends in the behaviour of the atmosphere, expressed in average figures for a large area. But, on the other hand, the number of time-steps necessary to advance computationally over a period of, say, fifty years, is vast.

And climate models have their own complexities. In the case of, say, a five-day forecast, reasonable results can be obtained by assuming that there will be no change in sea-surface temperatures over that period. But if you have to think in terms of decades, the changes in the thermal structure of the oceans will be the very driving force behind the processes involved. The behaviour of the ocean itself must be predicted, and the atmosphere and ocean must be 'coupled'—as the name CGCM implies.

And climate models must also accommodate variations in atmospheric composition. In the early days of climate modelling, attention was concentrated only on the possible consequences of increasing amounts of carbon dioxide in the global atmosphere, but other variable constituents, like nitrous oxide, methane and ozone, have been added to the recipe as the years went by.

THE APPARENT SIZE OF HEAVENLY BRIGHTNESS

5 February 2007 ∿

Have you admired, on recent evenings, the planet Venus shining brightly in the southwestern sky just after sunset? In case you wondered, its brightness at present, or more properly its 'magnitude', is somewhere around -3.9, and it will wax even brighter in the coming months, reaching a maximum of -4.6 by mid-July.

The brightness of any heavenly body depends not only on its absolute brightness—whether by virtue of intrinsic or reflected light—but also on its distance from the Earth. Everything else being equal, the farther away a star or planet is, the dimmer it will seem to be—just as in the case of a line of similar lamp-posts on a street, the nearer lights appear to shine more brightly than more distant ones.

Our system of magnitudes dates back to the second century BC, when the Greek astronomer Hipparcus divided the stars into six brightness groups; those stars first visible after sunset he designated as being of the 'first' magnitude, and the last to make an appearance, the very faintest stars visible to the human eye, were defined as of the sixth magnitude. Thus, like the handicap system used in golf, the system is inverted; the more brilliant performers are assigned the lower values on the scale, so that a star of magnitude 1.1 is brighter than a star of 2.2.

In the 1850s, Norman Podgson of the Radcliffe Observatory refined the whole system by defining a first magnitude star as 100 times brighter than the faintest star visible without a telescope, thereby introducing that which in mathematics is called a logarithmic scale. By this system it turns out that a first magnitude star is 2.5 times brighter than one of the second magnitude; six times brighter than one of the third; sixteen times brighter than a star of the fourth magnitude; forty times brighter than one of the fifth; and by definition, 100 times brighter than a star of the sixth magnitude.

Those with very good eyesight, particularly young people, may be able to distinguish stars of a lower magnitude than six. On the other hand, in light-polluted areas only magnitudes 3 or 4 may be distinguished by the human eye, and in all circumstances objects nearer to the horizon become harder to see because their light is attenuated by having to travel obliquely towards the observer a greater distance through the atmosphere. With good binoculars, however, stars of magnitude 8 or even 9 are visible in favourable conditions.

Unlike golfers' handicaps, the magnitude scale extends below zero to accommodate objects that are very bright indeed. Sirius, the Dog Star, for example, has a magnitude of about -1.5; Venus, as we have seen, is -4 or thereabouts, and even the Moon and the Sun can be accommodated on the scale at -12.6 and -26.8 respectively.

DIAMONDS IN THE SKY

6 February 2007 ∾

Captain Boyle in *Juno and the Paycock* was wont, often, to look up at the sky and ask himself: 'What is the stars, what is the stars?' Nowadays we can provide the answer with considerable confidence: they are huge concentrations of hot gases, exuding immense quantities of light, heat and other forms of radiation. They are quite different from the planets, which, by and large, shine only by the Sun's reflected light.

Looking at the night sky with the naked eye, about 6,000 twinkling stars are visible. Their twinkling, however, is not intrinsic; it is caused by variations in the density of our mobile atmosphere, which bends the light from the star and causes it to seem to tremble. Moreover, they are not all the same colour. Most of them vary from a sultry red to a steely blue, the difference being dictated by their temperature. Those with a reddish tinge are cooler than the blues, and the oranges and yellows, those like our Sun, lie somewhere in between.

The scintillation is closely related to the shimmering effect often seen when looking at the air over a hot surface on a very warm day. It occurs, not specifically because the air is hot, but because it is unevenly heated. A ray of light passing through air which is at a uniform temperature follows a perfectly straight path, but if it passes from warm air to cooler air, or vice versa, it changes direction ever so slightly at the boundary, a phenomenon called *refraction*.

Rays of light reaching our eyes from space pass through an atmosphere which is in constant motion, and through layers of air whose temperature and density are changing incessantly from second to second. These changes cause the rays of light from a star to follow an erratic and inconstant zigzag path to the eye. Indeed to catch the light from a given star, we have to look each instant in a slightly different direction. To put it another way, the star appears to 'dance' in the sky, rapidly changing its apparent position, albeit by a very small amount. The optical effect is that of 'twinkling'.

There are two other processes that enhance the twinkling effect. A blob of air warmer or cooler than its surroundings may some-times act temporarily as a lens, enlarging or diminishing the image of a distant star. Similarly, a volume of air may sometimes act as an optical prism, splitting the composite light from a star into its constituent colours; in the case of a 'white' star, for example, at one instant we may see the blue part of the spectrum, and at another the red. Then truly does the distant star appear as described nearly 200 years ago by Jane and Ann Taylor in the well-known nursery rhyme:

Up above the world so high,
Like a diamond in the sky!

DICKENS'S PATHOS STILTED BY THE WEATHER

7 February 2007 ～

' *Big whirls have little whirls*', wrote Lewis Fry Richardson, the father of numerical weather prediction,

that feed on their velocity;
And little whirls have lesser whirls,
And so on to viscosity.'

He had borrowed this template for a description of atmospheric vorticity from the nineteenth-century Oxford mathematician Augustus de Morgan, who was inspired by the concept of scale-invariance in astronomy to observe, by way of metaphor, that

Great fleas have little fleas
Upon their backs to bite 'em;
And little fleas have lesser fleas,
And so ad infinitum.

De Morgan in turn, of course, had lifted the format directly from Dean Jonathan Swift, but that is another story. The rather convoluted point of this whole rhyming argument is to illustrate that Augustus de Morgan had a clever way with words. Here he is, for example, on one of his illustrious contemporaries:

A splendid muse of fiction hath Charles Dickens,
But now and then just as the interest thickens
He stilts his pathos, and the reader sickens.

Charles Dickens was born in Portsmouth 195 years ago today on 7 February 1812, and during his boyhood he endured a spell of considerable misery when his father was imprisoned for non-payment of a debt. Dickens himself, at the age of twelve, worked in a blacking warehouse, and memories of this painful period inspired much of his later fiction. But it is also in his childhood that we may find the key to the dismal weather portrayed in many of his novels.

In mediaeval times, northern Europe had enjoyed a comparatively benign climate, but a sudden change took place around the middle of the sixteenth century: average temperatures dropped dramatically, and the period from about 1600 to 1850 was the coldest since the last Ice Age 10,000 years previously. The winters were long and very severe, and the summers cold and wet, as Europe emerged slowly from the rigours of this so-called 'Little Ice Age'. In particular, the decade from 1810 to 1820 was exceptionally cold and miserable in England, and the first nine winters of Charles Dickens's life would have been of a harshness quite unheard of in our own times. It seems likely that the corresponding childhood memories are reflected in his writing.

On a lighter note, we can infer that Charles Dickens would have approved of *Weather Eye*. A daily column on weather matters was one of the first regular features in the *Daily News*, a newspaper Dickens founded, and edited for a time, in 1846. The column was written by one James Glaisher of the Meteorological Department at Greenwich Observatory, and it seems to have been the first of its kind ever written for a daily newspaper.

THE SUBTLETIES OF SNOW

9 *February* 2007 ∽

Met Éireann, of late, has been falling into careless habits of accuracy. It has been my impression, anyway, that they predicted the onset and cessation of each of our winter storms for months on end with admirable precision. But then came yesterday:

Arrives the snow, and, driving o'er the fields,
Seems nowhere to alight; the whited air
Hides hills and woods, the river, and the heaven,
And veils the farmhouse at the garden's end.

Well that was what we were told was just about to happen, and in many parts, it must be said, it did; but in places there was much less snow than the radio had led us to believe there ought to be.

Meteorology, however, as we have often noted in this column, can never be exact and this is particularly true when it comes to fine-tuning between snow and rain. For one thing, no matter how many weather observations are available, it is never possible to describe the initial state of the atmosphere with exact precision. Moreover, the mathematical equations used by the computer to predict the weather are necessarily approximate, if for no other reason than that the calculations required must be within the capabilities of the machine.

And then how does the computer tell us that it is going to snow? Do red lights flash upon the consoles?—or does the machine simply display the landscape coloured white?—or is it a simple one-word message 'Snow!'?

Naturally, it is not as simple as that. The main output of the computer is a series of charts, which show pressure, temperature and precipitation patterns for selected future times. But even if these patterns are predicted with absolute precision, local topography and geographical features affect the amount and thickness of the cloud and the heaviness and persistence of any rain or snow. Knowledge and experience are required for the forecaster to translate the output of the machine into a meaningful interpretation of expected weather at a particular spot.

Snow is particularly difficult in this regard. In the simplest sense, we get snow instead of rain if the temperature is low enough—below about 3°C. But even if the ground-level temperature is predicted absolutely accurately, variations in the temperature several hundred metres above the ground also have bearing on the matter; the temperature may fall to 0°C and we still get rain. If the computer predicts that temperatures over the country tomorrow will be, say, 3°C or 4°C, then some places will get snow and others not, and it is a major challenge to identify which will be which. It may happen, too, that precipitation falling as rain in the afternoon may turn to snow with the approach of dusk, when the temperature near the ground drops below the crucial value. Or then again, it may well not.

| THE DUST-BOWL YEARS

10 *February* 2007 ∽

'Rain follows the plough' was a popular maxim in late nineteenth-century America. The phrase was first promulgated in a book written in 1881 by Charles Dana Wilber, a land speculator who claimed some expertise in climatology. His theory was that settlement and cultivation, especially the planting of trees, would transform the arid Great Plains of the United States into rich, fertile farmland endowed with copious amounts of rain.

Unusually abundant rainfall during the 1870s and early 1880s made these claims seem plausible, and within ten years nearly two million people had sunk their roots into the prairie soil. But when the wet years came to an end, the periodic droughts endured by the settlers reached their climax in the 'dust-bowl' years.

The immediate cause of the 1930s drought was an anomalous increase in the frequency of westerly winds in the middle latitudes during the six years from 1932 to 1937. Why this should have happened, even in retrospect, nobody quite knows; but happen it did, and in places like western Europe, whose rainfall is derived mainly from moisture transported from the west, the result was an increase in rainfall.

But in America most of the Great Plains lay for the duration in the giant rain-shadow of the Rocky Mountains. Any moisture the westerly winds might have gathered on their journey over the Pacific was extracted as rainfall before they crossed the mountains, and the hot, dry air that swept down into Colorado, Kansas and Oklahoma brought the worst drought experienced for generations.

Severe droughts occur regularly in this region every twenty years or so. On this occasion, however, the effects were exacerbated by the introduction during the previous decade of mechanised farming methods on a large scale. In previous droughts, dry winds affecting the native grasses of the area had produced a tough, dried-up mat that to a large extent protected the underlying soil. But now, once the drought had killed the crops, the soil that had been disturbed by the plough just blew away.

Strong winds brought severe duststorms that surged across the barren plains. People had to wear masks to protect themselves, and drive with the headlights on in daytime; dust piled up inside houses; schools and businesses were closed; and bereaved families were unable to bury their dead. Even the birds, it was said, were too afraid to fly.

When the climate finally returned to normal, much of the affected land was sensibly returned to pasture. In the meantime, however, the natural disaster of the drought had coincided with the economic hardship of the Depression, and thousands of ruined farmers and their families, epitomised by the Joads in

Steinbeck's classic *The Grapes of Wrath*, had to abandon their parched holdings and set out as 'Okies' along Highway 66 for the green promise of California to the west.

THE ADVENTURES OF THE MANY VALENTINES

14 *February* 2007 ∿

I t is a well-known fact, but one not always appreciated here in Ireland, that wolves are astute observers of the weather. When they howl and lurk near populated areas, it is a sure sign a storm is on the way—or as the Greek poet Aratus of Soli put it many years ago:

> *When through the dismal night the lone wolf howls,*
> *Or when at eve around the house he prowls,*
> *And, grown familiar, seeks to make his bed,*
> *Careless of man, in some outlying shed,*
> *Then mark!—ere thrice Aurora shall arise,*
> *A horrid storm will sweep the blackened skies.*

But it is also thanks to a wolf that we celebrate Valentine's Day the way we do. Romulus and Remus, the abandoned infant founders of the city of Rome, were suckled by a she-wolf in a cave that was subsequently called the Lupercal. When in due course the Romans held a festival of fertility in mid-February each year, participants gathered near the Lupercal, and the rituals were known as Lupercalia.

With the coming of Christianity, the somewhat raunchy Lupercalian celebrations came to be associated, quite coincidentally, with the feast of Valentine on 14 February, and by mediaeval times they had acquired a gentler, more romantic theme. Observers of the natural world, moreover, found in it a justification for their ardour, as Robert Herrick tells us:

Oft have I heard both youths and virgins say,
Birds choose their mates, and couple too, this day.

None of this amorous frivolity, however, has anything to do with Valentine himself, who, as far as we know, had no interest whatever in *affaires de coeur*. Indeed there are three possible St Valentines. One was a priest of Rome who, despite one or two spectacular miracles, was clubbed to death in AD 270 for harbouring persecuted Christians. A second Valentine, Bishop of Terni a few years later, also came to a bad end; he too was martyred for his adherence to the new religion. And there are rumours of a third Valentine, said to have passed away in Africa. One of these—presumably the priest of Rome—is said to have ended up in Dublin.

Fr John Spratt, a Carmelite priest attached to the Carmelite Church, Whitefriar Street, visited Rome in 1835. Fr Spratt was a brilliant preacher, and Pope Gregory XVI was so entranced by the persuasive eloquence of his Irish visitor that he gave the Carmelite a parting gift. It was not, as it might be nowadays, a case or two of wine, nor yet a plenary indulgence, but the body of St Valentine, rudely exhumed for the purpose from its putative resting-place in the St Hippolitus Cemetery. Valentine, we are told, was duly transported to Ireland and installed in 1836 with full saintly honours in Whitefriar Street, where he is venerated still.

HOW TO DO IT WITH MIRRORS

15 *February* 2007 〜

Mirrors are the most wonderful of toys. You can make kaleidoscopes with them; you can combine one with a lens to build an epidiascope; you can use them to admire your own reflection or to dazzle friends with irritating beams of sunlight; and if you were Archimedes, as we noted here a day or two ago, you could reduce the entire Roman fleet to ashes at the port of Syracuse.

Modern scientists, too, have a weakness for these toys. One brilliant wheeze was to use them to combat global warming, it being argued that a mirror on the ground would reduce the net terrestrial absorption of solar radiation. If, for example, an individual were to feel it sufficient only to neutralise his own personal share of the enhanced greenhouse effect—about one six-billionth of the total—he or she, it was calculated, would need some 150 square metres of mirror in the back garden. This, of course, applies only at the latitude of Ireland; conscientious residents of North Africa would need only 20 square metres of reflecting surface to discharge their duty to posterity.

Other more adventurous technocrats have suggested that the same objective could be achieved more tidily by deploying the array of mirrors on satellites in space, where they would serve as an extra-terrestrial parasol to shade us from unwanted radiation. Here the figures are intimidating; the cost of the energy required to launch the required acreage of mirror into space would amount to about 6 per cent of the world's total GNP over the next twenty years. Doing it with mirrors comes not cheap.

Another clever ploy was the proposal to alleviate the depressing darkness of the northern latitudes during the long winter months. The idea was to launch a fleet of giant mirrors into space, suitably orientated to reflect sunlight onto selected areas of darkness, producing thereby extended periods of artificial daylight. These mirrors, enthusiasts pointed out, could also be used to illuminate special events, and to provide light for search-and-rescue operations following disasters.

But perhaps the most imaginative suggestion for the potential use of mirrors has been the theory that they could be used to deflect rogue asteroids, should any be discovered heading on a collision course for planet Earth. The theory is that if danger threatens, a giant mirror would be launched into space in such a way that it would follow the asteroid along its track. As the two travelled side by side through space, sunlight focussed by the mirror onto the offending missile would evaporate loose silicate rocks or lumps of ice on the surface of the asteroid. This would result in tiny jets of gas that would act as thrusters, generating a small but steady push to nudge the cosmic body into a new, non-threatening orbit.

THE REDISCOVERY OF MAUNDER'S MINIMUM

21 *February* 2007 ∾

L ike Catullus, each of us shuffling off this mortal coil would like *meas esse aliquid putare nugas*—'to think my trifles worth a bit of notice'. Some, of course, received more notice than they might have thought ideal; Galileo, for example, might have preferred less interest in his work on the part of the Roman Inquisition. And Alfred Wegener might have wished that his peers were more receptive to his theories about continental drift; his bizarre and courageous notion was loudly rejected, and not without ridicule, during his lifetime. But Edward Maunder's discovery concerning sunspots left the scientific world bubbling over with complete indifference.

Maunder spent much of his adult life counting sunspots and browsing through records left by his predecessors down the centuries. Then in the early 1890s he noticed something very interesting indeed: between 1645 and 1715 hardly any spots at all had been reported by contemporary astronomers. Maunder published his findings in 1894, and again in 1922, on both occasions to resounding apathy; those who bothered to read his account at all assumed that he had got it wrong, or that the solar watchers all those years ago had been a little lax. It was not until nearly fifty years after Maunder's death in 1928 that someone decided seriously to investigate this strange alleged anomaly.

In the early 1970s, an American astronomer called Jack Eddy retraced the steps of Edward Maunder, and confirmed the existence of what has now come to be called the *Maunder Minimum*. Going even further, Eddy related the historical waxing and waning of sunspot activity to variations in the amount of the radioactive element carbon-14 to be found in contemporary timber. Then, using even older wood, he was able to extend the record of sunspot activity backwards, to periods long before the era of the telescope and in this way he identified two more anomalies. They were the so-called *Spoerer Minimum* from AD 1400 to 1510, and the *Mediaeval Maximum*—a period of unusually high sunspot activity—from AD 1100 until 1250.

Meteorologists in the intervening years have been quick to note that the two minima coincide nicely with the worst excesses of the 'Little Ice Age' in the northern hemisphere, and that the Mediaeval Maximum occurred at a time when we know our climate to have been unusually benign. The missing link, however, is any convincing mechanism by which sunspot activity might affect the climate of our planet.

Some have pointed out that cosmic radiation is modulated by the 11-year sunspot cycle, and suggest that this radiation may cause chemical changes which may affect the atmosphere's transparency—and hence the radiation balance of the planet. Most, however, remain unconvinced that sunspots are a significant factor in climatic change, or play any part in determining tomorrow's, or next century's, weather. But more on that anon.

AN ATTACK BY A FRENZIED SUN

22 February 2007 ◠

Sunspots, as we have seen, are dark blemishes on the surface of the Sun that increase and decrease in number over a regular cycle lasting about eleven years. But they are not just pretty pimples. They are areas of intense magnetic and electrical activity which sometimes culminate in a *solar flare*, shooting out great bursts of energy into interstellar space.

The energy from a solar flare arrives to Earth in stages. Electromagnetic waves in the form of X-rays, radio waves and the visible light which allows us to see the flare itself, take about eight minutes and give due warning of events to come. Cosmic ray particles are the next to arrive; they take about an hour. But the real 'magnetic storm', as it is called, reaches our planet about two days later.

Magnetic storm particles comprise ions and electrons, tiny charged entities herded together into a coherent stream by powerful electric fields. They have the potential to interfere with spacecraft, either by direct bombardment or by effecting changes in the density of the very tenuous high atmosphere. More spectacularly, as the charged particles interact with the magnetic forces associated with the Earth, strong bursts of electrical activity are induced in long metal objects on the surface—objects like railway lines, pipelines and electrical power lines. In effect, these lengths of metal act like dynamos to generate their own powerful surges of electricity.

Such storms are few and far between at present, since we are near the minimum of the solar sunspot cycle. They are most likely to occur, and to be at their most disruptive, when the number of sunspots is near maximum. In 1989, for example, a surge caused by a magnetic storm overloaded part of the electrical power grid in the United States, and caused a blackout to cascade through much of the northeastern United States and eastern Canada.

However, there is a bright side, quite literally, to magnetic storms: they facilitate more frequent and spectacular occurrences of *aurora borealis*, the northern lights. It was the eighteenth-century Swedish astronomer Anders Celsius, the same after whom the degrees of temperature are named, who first explained the origins of the aurora. He noticed that the northern lights seemed to follow the Earth's lines of magnetic force, and that they were more concentrated and more visible where these lines are most closely bunched together, near the Earth's geomagnetic poles. In effect, the planet's magnetic field acts like a giant TV cathode ray tube; the charged particles are channelled into beams, deflected towards the poles, and focussed onto the Earth's upper atmosphere, which acts in this respect like a fluorescent screen. This fluorescent luminosity provides the range of red, green, pink and blue lights that the aurorae emit to provide the brilliant displays that often decorate the northern sky.

FOGS AND SHRIEKS AROUND THE SKERRY

27 February 2007 ∽

Henry Wadsworth Longfellow's epic poem *The Song of Hiawatha* contains one of the most evocative pen-pictures of a red sunset to be found anywhere in English literature. It occurs when the crimson sky over Lake Gitche Gumee is pointed out to the young warrior by an old Indian woman called Nokomis:

> *Fiercely the red sun descending*
> *Burned his way along the heavens,*
> *Set the sky on fire behind him,*
> *As war-parties, when retreating,*
> *Burn the prairies on their war-trail.*

Longfellow was born in Portland, Maine, 200 years ago today, on 27 February 1807. In his late twenties he was appointed professor of French and Spanish at Harvard University, and by mid-century he was second only to Tennyson as the most popular poet of the English-speaking world. He retired from Harvard in 1854 to devote the remainder of his life to writing, and died in Cambridge, Massachusetts in 1882.

Now *Hiawatha*, as we know, is lyrical and sentimental stuff, but the gentle Henry could be menacing and macabre if it suited him. *The Skerry of Shrieks*, for example, describes an eleventh-century Easter party thrown by Olaf, King of Norway:

> *Now from all King Olaf's farms*
> *His men-at-arms*
> *Gathered on the Eve of Easter;*

In due course the revellers repair to bed, and the on-shore breeze, laden with moisture after its long passage over the Atlantic, cools as it drifts in over the chilly foreshore. Predictably, its moisture condenses into sea fog:

> *Pacing up and down the yard,*
> *   King Olaf's guard*
> *Saw the sea-mist slowly creeping*
> *O'er the sands, and up the hill,*
> *   Gathering still*
> *Round the house where they were sleeping.*

Soon there comes a rude awakening: Olaf's wily enemy, the druid Eyvind Kallda, uses this shroud of mist to launch a surprise attack upon the sleeping Norsemen. But Eyvind and his men are quickly overpowered, tied to a 'skerry' on the beach—an old Norse name for a rock—and nature is allowed to do the rest. Olaf and his guests

> *Silent sat and heard once more*
> *   The sullen roar*
> *Of the ocean tides returning.*

Finally Longfellow describes the gruesome end of the unfortunate Eyvind and his fellow villains:

> *Shrieks and cries of wild despair*
> *   Filled the air,*
> *Growing fainter as they listened;*
> *Then the bursting surge alone*
> *   Sounded on;*
> *Thus the sorcerers were christened!*

⏐ ROCK AROUND THE WEATHER

1 *March* 2007 ～

L et me purvey some meteorological statistics this morning which are bound to interest you. Dr Alan Robock of Rutgers University, New Jersey, has examined 465 songs performed over the years by rock singer Bob Dylan, and discovered that the word 'Sun' appears in 63 of them; 'wind' turns up in 55, 'rain' in 40, 'sky' in 36, 'cloud' in 23, 'storm' in 14, 'summer' in 12 and 'snow' in 11. Other weather-words occurring, but scoring less than 10, include 'hail', 'winter', 'lightning', 'thunder', 'flood' and, of course, 'weather' itself. It all goes to show that, as Dylan himself put it in *Subterranean Homesick Blues*, 'You don't need a weatherman to know which way the wind blows.'

Dylan became the voice of idealistic and disaffected youth in the early 1960s. His provocative and haunting lyrics in the new rock genre reflected the profound cultural and political changes sweeping through Western society at the time, and combined incisive social comment with brutal criticism of the perceived hypocrisy of the established order. But they had their weather, too.

In *Blowin' in the Wind*, for example, the title is tendered as the answer to nine questions posed in the body of the work. Of these nine, the majority are of no meteorological significance—like, for example, 'How many times must the cannon balls fly before they're forever banned?', or the rather esoteric query 'How many seas must a white dove sail before she sleeps in the sand?' But weatherpeople world-wide will empathise with the perennial frustration of 'How many times must a man look up before he can see the sky?' and some may well themselves have wrestled with the issue of 'How many years can a mountain exist before it's washed to the sea?'

A thunderstorm, on the other hand, is the dominant motif of *Chimes of Freedom*, its emotional impact invigorating a dramatic call to right the world. It contains an interesting description of the storm's departure:

> *Even though a cloud's white curtain in a far-off corner flared*
> *An' the hypnotic splattered mist was slowly lifting,*
> *Electric light still struck like arrows,*
> *Fired but for the ones*
> *Condemned to drift, or else be kept from drifting.*

Some, however, took Dylan's reforming zeal too far. 'You don't need a weatherman . . .' was adopted as an eponym by a radical and violent dissident student movement in the US in the early 1970s. 'The Weathermen' endorsed, and practiced, terrorism in their struggle against the establishment, and their aim was nothing less than to overthrow by force the American political structure of the day. Happily, by the late 1970s the egalitarian idealism of the Dylan era began to be replaced by the yuppie culture that was to predominate throughout the 1980s—and weathermen, once more, came to be identified only with their innocent isobars and fronts.

BALDER WITH A DASH OF SCIENCE

8 *March* 2007 ∽

S hould you be contemplating reading Michael Crichton's *State of Fear*, my advice would be: don't bother. But then *Weather Eye* is famously fussy and pretentious in its choice of reading matter; it is clear that hundreds of thousands of happy readers around the world have innocently found the book to be a thought-provoking and enjoyable experience. But science, art or literature it certainly is not.

Crichton's theme is climate change and global warming, and his basic plot is plausible enough. He envisages an ELF, an Environmental Liberation Front, whose members, despite—or perhaps because of—the intrinsic goodness of their cause, are willing in its furtherance to go to the same extremes of terror and intimidation as those alleged of some members of the Animal Liberation Front.

With an important climate conference in the offing, this faceless ELF sets out to give global warming a little technical assistance; it arranges for a number of high-profile environmental catastrophes to jog the delegates towards the right conclusions. These devilish ploys are in the same genre as the concept behind Crichton's other successful book, *Jurassic Park*, which envisaged the reconstruction of the dinosaurs from tiny preserved samples of their DNA. The techniques involved here, too, are not scientifically possible at present, but there is a slight and entertaining possibility that some day, perhaps, they might be.

This time our villains have in hand three major projects: by a series of carefully synchronised explosions in Antarctica, they plan to set adrift a massive iceberg. They electrically enhance an already powerful thunderstorm to causes flash floods and serious loss of life; and they use more detonations to cause an underwater landslide near New Guinea, which they hope will send tsunamis surging towards America.

Our heroes' task, of course, is to foil these rotten plots. This they do by hectic voyages in a private jet, giving each other lengthy lectures about climate change *en route*. They are climate sceptics to a man, of course—and woman—and although Crichton has researched his subject very thoroughly, and quotes scientific references for all the views his characters so ardently espouse, these references, though genuine, are highly selective and paint a picture far removed from the current climate-change consensus.

Not only is our heroes' science unbelievable, but they themselves are even more incredible. Some are indistinguishable from one another; most seem to have no reason whatever for being where they are at any particular time; and any unfortunate passer-by who says anything about global warming is treated to a rude and arrogant tirade of several pages to persuade him, or her, of the total error of such views.

Otherwise the book is good clean fun. But this reviewer, seeing that one of the characters was a Mr Balder, was disappointed not to find another by the name of Dash.

DO PARACHUTES REALLY WORK?

10 *March* 2007 ∼

A week or two ago, the dangers of inferring a cause-and-effect relationship between two phenomena which appear for a time to vary in a similar way was illustrated in this column by reference to a satirical scientific paper which examined birth-rates in various parts of Germany. It purported to show 'a significant correlation between the increase in the stork population and the increase in deliveries outside city hospitals'.

But the reverse also applies. A reader of this column has kindly drawn my attention to a paper in a 2003 issue of the *British Medical Journal* intended to illustrate that the absence of rigorous trials and studies to prove a hypothesis does not necessarily imply that the hypothesis itself is incorrect: merely that it has not been proved.

The article, 'Parachute use to prevent death and major trauma related to gravitational challenge', was co-authored by Professor Gordon Smith and Dr Jill Pell. The parachute, they point out, is 'used in recreational, voluntary sector, and military settings to reduce the risk of orthopaedic, head, and soft tissue injury after gravitational challenge, typically in the context of jumping from an aircraft.'

But the perception that parachutes are a successful intervention, they say, is based largely on anecdotal evidence. 'If failure to use a parachute were associated with 100 per cent mortality, then any survival associated with its use might be considered evidence of effectiveness. However, an adverse outcome after free-fall is by no means inevitable. Survival has been reported after gravitation challenges of more than 10,000 metres (33,000 feet). In addition, the use of parachutes is itself associated with morbidity and mortality, often due to failure of the intervention.'

Moreover, the samples in the observational data may be biased, falsely enhancing the apparent protective effect of wearing a parachute: 'Individuals jumping from aircraft without the help of a parachute are likely to have a high prevalence of pre-existing psychiatric morbidity. Individuals who use parachutes, on the other hand, are likely to have less psychiatric morbidity and may also differ in key demographic factors, such as income and cigarette use.'

Studies are required, therefore, according to the authors, to calculate the balance of risks and benefits of parachute use. 'It is often said that doctors are interfering monsters obsessed with disease and power, who will not be satisfied until they control every aspect of our lives (Ref: *Journal of Social Science*, pick a volume). The widespread use of the parachute may just be another example of doctors' obsession with disease prevention and their misplaced belief in unproved technology to provide effective protection against occasional adverse events.'

The authors conclude: 'Everyone might benefit if the most radical protagonists of evidence-based medicine organised, and participated in, a double blind, randomised, placebo controlled, crossover trial of the parachute.' Alternatively, they say, common sense might occasionally be permitted to prevail.

NICE—BUT NOT THAT NICE

15 March 2007 ∾

'The astringent mildness of our dominant maritime-polar air is one of the few unalterable comforts we possess, however much we may personally deplore its harassing benevolence. The air that reaches us, in the words of Shakespeare's Ariel,

> *... suffers a sea change*
> *Into something rich and strange.*'

Nowhere are these words of Professor Gordon Manley, the eminent early twentieth-century English climatologist, more apposite than in the extreme southwest of Ireland.

The region is a zone of contrasts. The benefits of the warm prevailing winds, for example, are not unmitigated, since the mild winter temperatures are balanced by frequent gales, which have a negative effect on flora. In exposed places, particularly out on the peninsulas, growth is stunted by the strong winds and by the salt they carry from the nearby sea. The growth of buds and branches on the trees is checked, and where they survive at all they lean inland and become deformed and wind-shorn. And the mountains, heavily marked by glacial scars, are thinly clad in peat and heather. This almost barren treelessness is only softened by the profusion of gorse, and by the fuchsia hedges which are a special characteristic of the rugged landscape during the summer months.

But where there is shelter from this wind and its pervasive salt, the extreme mildness of the climate and the abundant moisture produce spectacular effects. Here and there are pockets of extreme fecundity, where the vegetation is of almost tropical luxuriance. Shelter has allowed the famed Killarney oaks to flourish, and given us the luxuriant gardens of Inacullen, Garinish, Rossdahan and Glanleam. Where there is shelter, Cork and Kerry—here and there—might be 'this other Eden, demi-paradise'.

It is sometimes fondly said that these sheltered parts of the region enjoy an almost Mediterranean climate, but such assertions have to be judged by three criteria. Firstly, except for an occasional thunderstorm, virtually all the rain in a Mediterranean climate falls in the winter half of the year.

Secondly, the summers are hot, dry and almost cloudless, and the mild Mediterranean winters are far less cloudy than those of northern Europe; week-long palls of cloud, not uncommon here, are rarely seen. And the third characteristic is the great amount of sunshine, well in excess of 2,000 hours per year.

Other areas of the world qualify by these criteria to have their climates described as 'Mediterranean'. Notable among them are the coast of California in the United States, the southwest corner of Australia, the area around Cape Town in South Africa, and parts of central Chile. But evidence for including Cork and Kerry on the list is not persuasive. Our summers are dull, damp and cool by comparison with Nice or Cannes, and even by their sunshine statistics alone, all parts of Ireland are instantly disqualified.

A CITY DIES OF THIRST

16 March 2007 ∿

Do you remember the opening lines which run like this?

It seems no work of Man's creative hand,
By labour wrought as wavering fancy planned;
But from the rock as if by magic grown,
Eternal, silent, beautiful, alone!

Probably not. And if I tell you that the author was a nineteenth-century Oxford don called John Burgon, who late in life became the Dean of Chichester and is still quoted extensively as an expert on the Sacred Scriptures, you will probably be none the wiser. But if I give you the last two lines of the sonnet, I have no doubt that memories will come flooding back:

Match me such marvel save in Eastern clime,
A rose-red city half as old as time.

The city, as we know, is Petra, the ruins of which lie in modern Jordan about half-way between the Dead Sea and the Gulf of Aqaba. The site is famous for having a great many magnificent buildings carved into solid sandstone, although Dean Burgon never visited, and the rocks of Petra are in fact of many different hues, few of which could plausibly be said to be 'rose-red'.

Petra was the capital of the Nabateans, Arabic-speaking Semites, from around the fourth century BC, and was built at the intersection of the many caravan routes that criss-crossed the Middle East around that time. Walled in by towering rocks, the city had all the advantages of a natural fortress, but its success was also largely due to the skill of its inhabitants in husbanding the water of the region.

It was in this particular wilderness, according to the Bible, that Moses struck a rock from which water gushed to meet the Israelites' needs; indeed the hinterland of Petra is still known as the Valley of Moses, *Wadi Musa*. In Nabatean times, this water continued to be plentiful, and the city appears to have had a Mediterranean type of climate, with abundant rainfall during the winter months. It is known from archaeological remains that the Nabateans controlled and stored the water by means of conduits, dams and cisterns, thereby ensuring a plentiful supply even during the long, hot, dry summers to allow the cultivation of fields and gardens and the growth of a civilisation where agriculture flourished.

Petra declined rapidly during the period of Roman rule. This was largely because the trade routes changed significantly, but hydrologists point out that present-day Petra is an arid, dusty, desert place, and they contrast this with the ample rainfall that was clearly available during the city's prime. It seems likely that a gradual change in climate, bringing prolonged and frequent droughts beyond the capacity of the ingeniously constructed water systems, may well have been a factor in the demise of that 'rose-red city half as old as time'.

WHAT WERE THE SIGNS THAT GOETHE SAW?

22 *March* 2007 〜

Johann Wolfgang von Goethe, poet, lawyer, painter, government minister and polymath *extraordinaire*, was born in Frankfurt in 1749. He graduated in law from the University of Strasbourg, but quickly abandoned that profession, devoting himself instead—if the lawyers among us will forgive me—to more intellectual pursuits. He settled in Weimar, where he became a trusted confidant of the ruling Duke, under whom he achieved high government office. Apart from a brief sojourn in Italy, Goethe lived in Weimar for the remainder of his life, and it was there he died, 175 years ago today, on 22 March 1832.

Goethe, as we know, is best remembered for his version of *Faust* and for many other literary works, but he also dabbled in the world of science, producing respected texts on biology and optics. His biology has proved to be enduring; his views on optics, on the other hand—largely a refutation of classical Newtonian theories—have not survived the test of time, although his book *Farbenlehre* conveys interesting asides on our perceptions of the different colours.

Much of our information about the personal side of Goethe's life comes from the memoir *Gespraeche mit Goethe*, published in 1836 by Goethe's personal secretary throughout the last nine years of his life, one Johann Peter Eckermann. Translated as *Conversations with Goethe*, it contains an interesting episode concerning Goethe's apocryphal ability to sense distant earthquakes.

'Once there was a call in the middle of the night,' writes Eckermann, 'and when I went to the room where he lay, I found him observing the sky. He said to me, "This is a significant moment; at this moment we are either having an earthquake, or will have one soon."'

What exactly Goethe saw, we do not know. Over the centuries there have been persistent rumours of 'earthquake lights' as precursors of a seismic tremor, and these reports have continued into recent times. Residents of Tangshan in China, for example, were awakened one night in 1976 by bright flashes in the sky, and the next night an earthquake registering 7.8 on Richter's scale killed 240,000 people and destroyed the city. And in 1999, pictures of balls of light apparently floating in the air were broadcast on Turkish television, purportedly filmed on the night before the devastating Izmit earthquake in which 15,000 people lost their lives.

One might speculate that Goethe observed or imagined some putative seismic precursor of this kind. In any event, Eckermann continues: 'The next day my master related his observations at court, at which one woman whispered in her neighbour's ear, "Listen, Goethe's mad!" But the Duke and the other men believed in Goethe, and it was soon revealed that what he had seen was correct, because after a few weeks the news arrived that on that very night a part of Messina had been destroyed by earthquake.'

THE QUIRKS AND TWISTS OF TIME

24 March 2007 ～

There is a strange asymmetry about implementing 'summer time'. One might innocently think that 'winter time' ought to apply in a symmetrical way around the winter solstice, but this is not the case. In 2006, for example, summer time ended on 29 October, fifty-three days before the solstice, whereas the 'spring forward' of the clocks does not take place until tonight, ninety-four days after the solstice. This is to ensure, apparently, that the mornings are not too dark in late February and early March.

The problem arises because the Sun is not as good at keeping time as one might think. If you measure the time directly by the Sun, the days, rather surprisingly, turn out to differ significantly in length in the course of the year, one of the reasons being that the Earth moves faster in its orbit at certain times than it does at others. As a consequence, in late October the Sun reaches the noon position about 15 minutes earlier than it would if the days were all of equal length; in early February, the Sun each day is 15 minutes late. To order our affairs without this complication, in everyday life we *pretend* that the days are all exactly twenty-four hours long; we use 'mean time', a term familiar to us in the guise of the Greenwich Mean Time we use throughout the winter months.

Now the benefits of summer time are gained in the longer evenings. The price we pay for this amenity—and a sensitive one in the context of road safety and farming practice—is in the dark mornings and the apparently late sunrise in the spring and late autumn. And because we use 'mean time', there is an asymmetry about the winter solstice in this lateness of the morning sunrise.

When it comes to adjusting our clocks, no one has a major problem with the late October date. Indeed, the change in the clocks around that time causes sunrise at the Greenwich meridian, the relevant benchmark for our timezone, to revert from around 0745 to 0645, which usefully delays the onset of dark winter mornings. But fifty-three days after the winter solstice—around 13 February and as long after the solstice as late October is before—sunrise takes place around 0720—later than you might think, because of the mean time anomaly discussed above. If we were to change the clocks in mid-February, sunrise would *advance* by an hour to 0820, giving us a much later sunrise than any we experience at present. But by waiting until late March, when sunrise, winter time, is around 0550, advancing the clocks brings it to 0650—which, not entirely by coincidence, corresponds very closely with the time of sunrise just after the clocks had gone back the previous October.

'There's currants for cakes,' as they used to say, 'and raisins for everything.'

THE CURSE OF THE BRINDLED COW

29 March 2007 ～

Sometimes, at this time of year, winter has a last, final fling and a harshness more appropriate to the dismal depths of February unexpectedly overwhelms the would-be gentleness of the days that lead us into April. Francis Ledwidge describes it as a time

> *. . . when loud March blows*
> *Thro' slanting snows her fanfare shrill,*
> *Blowing to flame the golden cup*
> *Of many an upset daffodil.*

The reason for this temporary but perennial harshness of the elements can be traced to an ancient cow. Indeed, later on in the same poem, *Lament for Thomas MacDonagh*, Ledwidge makes reference to such an animal:

> *But when the Dark Cow leaves the moor,*
> *And pastures poor with greedy weeds,*
> *Perhaps he'll hear her low at morn*
> *Lifting her horn in pleasant meads.*

But the beast responsible for the cold snap that frequently occurs around the end of March is a cow of a rather different colour: the *Bó Riabhach*, or brindled cow, of ancient Irish legend. March, it seems, was once rather shorter than it is now, and many centuries ago, at the beginning of the old April, the brindled cow began to grumble to her bovine friends about the harshness of the previous month. March at first took very little notice but, as the crescendo of complaint increased, understandably took pique, and resolved to teach the cow a lesson. March duly borrowed three days from April, and made them so cold and miserable that the *Bó Riabhach* perished with this unprecedented harshness of the elements. Ever since, this sudden cold snap has been called *Laethanta na Bó Riabhaí*.

The volatility of Irish weather in late March and early April is also implicit in the notion of the 'blackthorn winter'. Ancient wisdom has it that there are generally a few mild days towards the end of March which bring the blackthorn bushes into bloom. But these, it seems, are invariably followed by a cold harsh period— traditionally known as the blackthorn winter.

Many other European countries have their own version of the so-called 'borrowed days'. In England, for example, there are no bovine complications. The story runs simply:

March borrows time from April,
Three days—and they are ill:
The first is frost, the second snow,
And the third is cold as it can blow.

And in Scotland, this unwelcome cold snap is sometimes called the 'Peewit's Pinch'. The peewit is a kind of plover or lapwing, onomatopoeically named from its distinctive cry, and it often feels the 'pinch' when a few days of cold winds and frosty weather come upon it unexpectedly, just as it is about to build its nest upon the Scottish moors.

THE ELEMENTS OF A
CONSPIRACY REVEALED

30 *March* 2007 ∿

'All professions', said George Bernard Shaw, 'are conspiracies against the laity.' Several recent books and television programmes about global warming accuse meteorologists of being more prone than others to this grievous fault: we deny it utterly, of course, most utterly, but will concede that from time to time the words we use may well confuse the opposition.

One such is radiation. The process is invoked to explain a frosty night, or quoted as the mechanism by which weather radar works; or it may be worked into a complex explanation of the depletion of the ozone layer or, *a fortiori*, global warming.

The phenomenon is *electromagnetic radiation*, a process by which energy 'radiates' or flows outwards from a body. It is propagated in the form of electromagnetic waves, so called because they may be thought of as a rhythmic increase and decrease in the strength of an electric or a magnetic field, analogous to the rise and fall in the level of a water-surface.

Electromagnetic waves vary in length from tiny fractions of a metre to well over a kilometre, and they have very different characteristics depending on their wavelength. The very shortest are the gamma rays and X-rays, each much less than a micrometre long. They have important industrial and medical uses because of their great penetrative qualities, but for this same reason they are potentially harmful to us human beings.

Next, in order of increasing wavelength, come ultraviolet, visible light and infra-red waves, which together make up most of the sector of the spectrum occupied by the radiation from the

Sun. Ultraviolet waves, too, are harmful, but we are largely protected from them by an absorbent layer of ozone in the upper atmosphere. In the case of visible light, the different wavelengths allow us to distinguish separate colours, the longest—at about seven-tenths of a micrometre—being those which our eyes detect and interpret as the colour red.

Radiation whose wavelength is slightly longer than red, ranging from a millionth to a thousandth of a metre, is not possible for the human eye to see. We call it *infra-red*, and we can sometimes sense it in the form of heat. The energy lost to space by the Earth, and whose loss causes a dramatic drop in temperature on clear, frosty nights, is radiated in the infra-red portion of the spectrum.

Artificially generated electromagnetic waves of greater wavelength, some centimetres long and known as microwaves, can be used for cooking and for weather radar. And finally, radio waves, used in telecommunications, are anything from a metre to a kilometre in length.

All these are essentially the same phenomenon. They travel at the same speed, the so-called 'speed of light' or 186,000 miles per second in a vacuum. It is only in their wavelengths, and in their effects and capabilities, that they are different.

FIXATED BY THE CLOUDS

31 March 2007 ～

Cumulus clouds are much beloved of landscape painters. Their attraction lies, no doubt, not just in the interesting shapes that they assume, but also in the fact that cumulus clouds typically occur on bright, airy, sunny afternoons, when the colours of the countryside are shown to best advantage. Since there are gaps between the cumuli, the artist can often capture a landscape pleasingly dappled with passing shadows. By contrast, clouds formed at higher altitudes are generally layered, making the landscape dull and unstimulating, perhaps, from an artist's point of view.

John Constable was an expert on these matters. Born in Suffolk in 1776 to a wealthy milling family, he was fortunate enough to be of independent means. He recognised the weather as an integral part of painting, and his skies in particular were the product of a lifetime's study. In the summers of 1821 and 1822, for example, he produced a series of detailed oil sketches devoted exclusively to clouds, some fifty of which remain extant; each is carefully inscribed on the back with the time of day, the direction of the wind, and other memoranda relevant to the weather at the time.

Constable's fixation with clouds turned landscape painting on its head. 'The landscape painter who does not make his skies a very material part of his compositions', he wrote, 'neglects to avail himself of one of his greatest aids.' Practising what he preached, instead of allowing the landscape itself to dominate a picture as it had done for many of his predecessors, Constable made the sky, as he put it, 'the keynote, the standard of scale, and the chief organ of sentiment' of all his later works.

Because he learned his craft so carefully in this respect, Constable is viewed approvingly by meteorologists. His clouds are seen as scientifically accurate and as excelling in their dynamic quality, incorporating always a subtle intimation of continuing change, and he was one of the earliest painters to recognise the influence of topography on the shapes of clouds. It may be that his expertise on skies owes something to his known familiarity with the work of Luke Howard, a contemporary with an interest in meteorology who devised the scientific system of cloud classification still in use today.

By contrast, in other meteorological respects Constable's *oeuvre* is strangely limited in scope. He confined himself to simple rustic themes, mostly in the southeast of England and depicted always in the summer season; snow, ice or frost were totally taboo. Moreover—in contrast to Turner, for example—visibility in Constable's paintings is nearly always good; he rarely painted mist or fog, and almost invariably, his horizons can be clearly seen.

John Constable died suddenly 170 years ago today, on 31 March 1837.

THE FICKLE CHARMS
OF APRIL

2 *April* 2007 ∿

No one is quite sure where April got its name. Some believe it is related to the newly evident reproductive powers of Nature, and that it derives from Aphrodite, the Greek goddess of fertility and love. Others blame the Romans, who, they claim, saw this month as the gentle angel of maturing spring; to the Romans, April was the month that 'opened up' the year, and restored to Nature all its fruitfulness. '*Omnia aperit*', they would say of it: 'It opens all things,' and hence its name of April.

Henry Wadsworth Longfellow seemed to be of the latter persuasion. In any event he wrote of April:

I open wide the portals of the Spring
To welcome the procession of the flowers,
With their gay banners, and the birds that sing
Their song of songs from their aerial towers.

Whichever side you choose, the month of April is by any standards a marked improvement on its predecessors. Its brightness is not an illusion brought about by the changing of the clock, nor even entirely by the longer days—although of course both these factors play a part; the average duration of bright sunshine in April exceeds that of March by a full hour and a half.

We can expect less windy weather, since, on average, there are significantly fewer gales in April than there are in the preceding months. Moreover, this is statistically the driest month of the year in most parts of Ireland. But April showers are a reality despite the low rainfall totals. Noticeable rain—more than 1 millimetre—falls on eleven or twelve days of the month, and what you might call a 'decent fall of rain', more than 5 millimetres, occurs, on average, on about three or four of April's thirty days.

When the forecaster speaks of 'normal' temperatures this month, he or she will be thinking in terms of about 12°C to 14°C in the afternoons; indeed, once every four or five years or thereabouts, the afternoon temperature on an April day creeps above 20°C. Air frost can still be expected to occur inland, however, on three or four days of the month, and ground frost on about one day in three.

Although it has the reputation of being a very gentle month, with its longer days, its daffodils and its emerging greenery all calculated to induce in the unwary a premature complacency about advancing Spring, it is a mistake to be seduced too much by April's fickle charms. It can be something of a charlatan; ninety years ago, for example, in the early days of April 1917, there was widespread disruption throughout the country due to severe snowstorms. Indeed, the warning about casting clouts in May has its counterpart to suit the current month:

Till April is dead
Change not a thread.

THE COMPLEXITIES OF CALCULATING EASTER

7 April 2007 ～

E aster takes its name from *Eostre*, the Teutonic goddess of the dawn, whose heathen festival was held around the vernal equinox. Our Christian Easter is related to the Jewish feast of Passover, which in turn commemorates the 'great deliverance'—when the destroying angel 'passed over the houses of the children of Israel in Egypt'. And its significance for Christendom, as we know, is its association with the Resurrection.

Very early Christians simply followed the Jewish calendar, but as the new religion became established, there followed a desire for independence and many different methods were employed to arrive at the appropriate date. These diverse views were reconciled, or so it was thought, at the Council of Nicaea in AD 325 when it was decreed that Easter should be the first Sunday after the first full Moon after the vernal equinox. Moreover, to make matters easier in an age when local astronomers were hard to come by, the vernal equinox could be taken as 21 March, although it does not always fall upon that day. Things could not be simpler.

Well, perhaps they could! Confusion continued, much of it centred about the issue of when exactly the Moon could be described as being full. Eventually there emerged two trains of thought: the East favoured the calculations of Victorius of Aquitaine, while those of the Roman bailiwick preferred those of Dionysius Exiguus—the sobriquet 'the Little', we are told, being a reference to his demeanour of humility rather than a statement on his stature. When the dates differed, each side had its miracles—like a font which filled spontaneously with holy water on Easter Sunday morning, or a miraculous spring which flowed only on that day—to prove conclusively that its calculations were correct. In due course, however, the system proposed by Dionysius came to be universally accepted.

Well, almost! Irish monks, who were very influential throughout Scotland and the north of England, had their own way of calculating Easter's date which was different from any of the continental systems, and to which they clung with a famously vociferous tenacity. This divisive issue was only finally decided at the Council of Whitby in Yorkshire in 664; King Oswy, the local potentate, greatly influenced by Augustine of the See of Canterbury, leaned predictably towards Rome and the way of calculating Easter was firmly settled.

Well, nearly! When most of western Europe changed from the Julian to the Gregorian calendar in 1582, the Eastern Churches refused to have anything to do with it. They retained the Julian system for their Easter calculations—including a vernal equinox fixed on 21 March, Old Style, which falls in early April by our present reckoning. It is for this reason that the Greek Orthodox and the Roman Easters often fall on different dates—although, as it happens, this year they coincide.

THE ODDS AGAINST THE BOYS

12 *April* 2007 ∽

Nature is something of a gender-bender. Boys, at birth, outnumber girls by a few percentage points, but Nature has cleverly, and rather unfairly some might say, compensated for this imbalance by arranging for men, on average, to die sooner than their womenfolk, and so it evens out. Indeed the imbalance between the sexes begins to be redressed at an even earlier stage in life, in that males, it seems, have a significantly higher mortality rate in the womb and around the time of birth.

There are also seasonal aspects to this battle of the sexes. According to research published some time ago in the journal *Human Reproduction*, significantly more boys, relative to the total number of births, are conceived between September and November, while fewer are conceived between March and May. Italian gynaecologist Angelo Cagnacci and his colleagues looked at more than 14,000 births over a six-year period. The normal gender ratio is 511 males to 489 females per 1,000 conceptions, but the study revealed that 535 males were conceived during the favourable months mentioned above, compared to 465 females; in the least favourable months only 487 males were conceived, compared to 513 females.

Cagnacci and his friends surmise that Nature may compensate for the fact that male foetuses and new-borns are more fragile than their female counterparts by ensuring that more boys are born at optimal times of the year. Boys seem to be favoured during the most advantageous reproductive conditions, since babies conceived between September and November will have the advantage of being born in the warm summer months.

Moreover, boys born during the winter months may sometimes have a further threat to their longevity: they seem to have a greater risk of growing fat. A few years ago a team of British scientists looked at 1,750 men and women who had been born in Hertfordshire between 1920 and 1930 and who had lived there all their lives. In addition, they classified the Hertfordshire winters from 1920 to 1930 inclusive as either 'mild' or 'cold'. They found that men born during a cold winter were significantly more likely to become clinically obese later in life than those born in mild winters.

The research suggests that although lifestyle and genetics clearly play an important part in whether or not people put on weight, the influence of environment on the foetus may also have an impact. The seasonal factor, moreover, is predominantly focussed on the male; a similar 'cold winter' propensity to obesity, while just about detectable in the case of women, was decidedly less pronounced. Other research suggests that it is the *relative* coldness that matters, rather than the absolute degree thereof, consistent with another observed phenomenon, that the children of people who emigrate from hot to colder climates are more prone to obesity than the population at large.

HANDEL ON HAILSTONES

13 *April* 2007 ∼

George Friedrich Handel has been described as 'the man who set the Bible to music'. He took to oratorios late in life, only after his earlier adventures as an opera composer had brought him mixed returns; he had some successes, but he had even more failures, and was twice bankrupt as a result of over-lavish and commercially unsuccessful opera ventures. It was when he was in his-mid fifties, in debt, and in poor health after suffering a stroke, that he turned to the new *genre* and thereby found his *métier*.

Messiah is Handel's best known work in this form, and it was first sung in Neale's Music Hall, Fishamble Street, Dublin, 265 years ago today, on 13 April 1742. Handel had come to Dublin in November the previous year, invited by the Viceroy, the Duke of Devonshire, 'to compose something special in aid of Dublin's sick'. The premiere was performed by the combined choirs of Dublin's two cathedrals—Christ Church and St Patrick's—and the *Dublin Journal* reported the performance as 'superior to anything of the kind heard before in this Kingdom'. But to some, *Messiah* is disappointing for its dearth of meteorology; indeed, the only obvious reference to the weather is the joyous prediction in the concluding *Hallelujah Chorus* that 'He shall rain for ever and ever'.

A less popular Handel oratorio is in this respect of greater interest. *Israel in Egypt*, written in 1739, is concerned with the plight of the Israelites in their Egyptian bondage, and describes, *inter alia*, a sequence of plagues sent by God to trouble the Egyptian Pharaoh.

There were ten, altogether. The first was of water turning into blood; then followed three plagues of pests—frogs, mosquitoes, and then flies. A disease affecting livestock was followed by an epidemic of boils which afflicted man and beast, after which there was a lethal storm of hailstones, and then locusts, darkness, and the death of first-born children.

The seventh plague, as immortalised by Handel, includes one of the most powerful musical depictions of a hailstorm to be found anywhere in classical music. The aria 'He Gave Them Hailstones' follows the Biblical text from Exodus: 'The Lord sent thunder and hail, and lightning running along the ground, and rained hail upon the land of Egypt. And it was of so great bigness as never before was seen in the land of Egypt since that nation was founded; and the hail destroyed all things that were in the fields, both man and beast; and it broke every tree of the country.'

It is a dramatic example of the Almighty using the weather as a tool to change the course of history; indeed, it is an interesting thought that of the ten plagues sent by God around that time, no less than eight were related in some way to meteorology.

| APOLOGIA PRO JOCO SUO

16 *April* 2007 ∾

The late Herr Handel caused a little trouble. I was reminded over the weekend of Byron's assertion in *English Bards and Scotch Reviewers* that:

> *A man must serve his time to every trade*
> *Save censure—critics all are ready made.*
> *Take hackney'd jokes from Miller, got by rote,*
> *With just enough of learning to misquote.*

Apart from Miller—a London publisher not always receptive to Lord Byron's *oeuvre*—the last two lines describe precisely my approach to *Weather Eye* on Friday. I did not expect to be taken seriously when I recalled that the only obvious reference to weather in *Messiah* is 'the joyous prediction in the *Hallelujah Chorus* that 'He shall rain for ever and ever'. It was, *à la* Byron, a hackney'd joke, displaying 'just enough of learning to misquote'. A significant number of readers, however, e-mailed to register their gentle tuts. My apologies to those who were misled; perhaps in future I should place 'ha-ha' in brackets after sallies into wit, or 'lol', or simply several '!!!!'s.

Let me tell you of something else I recalled, about which, in the current circumstances, you will almost certainly conclude that I am joking. Back in 1994, a proposal to sell water from Lough Corrib to wealthy clients in the Middle East gained significant momentum. The idea was that some 10 million gallons of water per day would be removed from the lake, pumped into giant tankers, and transported to parts of the world where that commodity is in very short supply. The project never reached fruition; perhaps in Riyadh and Kuwait at present the people offer thanks for their deliverance.

Needless to say, this was not the first ambitious scheme dreamed up to solve the water problems in arid regions of the world. In the 1970s it was suggested that large icebergs might be towed southwards from the polar region to be anchored, for example, off Saudi Arabia, and the water from them harvested as necessary. Best estimates at the time were that only 50 per cent of an iceberg would be lost, *en route*, by melting.

But there were other snags. First there was the physical problem of towing a giant lump of ice; an iceberg of suitable size would be too heavy to be handled by even the most powerful tugs. Moreover, the process of attaching a cable—assuming one knew how to do it—would be a hazardous one, since icebergs have a nasty habit of suddenly toppling over when their centre of gravity shifts position as bits of ice break off.

And then there would be problems at the destination. You cannot berth a giant iceberg neatly by the quayside, since nine-tenths lies underneath the surface; processing the ice some distance from the shore to obtain the required water would pose a major challenge.

THOUGHTS ON A HAZY BEACH

20 *April* 2007 ∿

'Queer are the ways of a man I know:' wrote Thomas Hardy:

> *He comes and stands*
> *In a careworn craze,*
> *And looks at the sands*
> *And the seaward haze. . .*
> *And what does he see when he gazes so?*

We can assume, I suppose, that Hardy's acquaintance would see the horizon, and taking him to be a man of average height standing on the beach, his horizon would be some 5 kilometres away. If, on the other hand, this worried eccentric was on a cliff, say, 30 metres high, his horizon would recede to 20 kilometres or thereabouts. And were he standing on a hill beside the sea, his horizon might well be more than 50 kilometres distant.

Of course, it is sometimes possible to see considerably farther than 50 kilometres if we are looking at a very large object projecting some distance above where our horizon ought to be— like, for example, at a distant mountain. Indeed, if the air were perfectly clear, visibility would be in the region of 250 kilometres.

But the air is never really clear. Rays of light heading towards an observer from some distant object undergo a continuous attenuation caused by impurities or moisture particles suspended in the atmosphere, causing visibility to be reduced by either mist or haze.

Haze is a reduction in visibility caused by tiny *solid* particles suspended in the air, little motes of dust or smoke, that may come from a great variety of sources; they might be desert sands or particles of arid soil advected from afar, ash from some volcanic eruption in a distant land or, more commonly in this part of the world, local domestic or industrial pollution.

The proliferation or otherwise of these haze particles is highly dependent on local weather conditions. If there is little upward or downward motion of the atmosphere, as happens in an anti-cyclone, the obscuring particles remain in the stratified layers of air near the ground and reduce visibility significantly. Where, however, there is significant vertical movement of the air, as in the unstable, showery airmass behind a cold front, the particles are dispersed throughout a very deep layer of the atmosphere and visibility is good.

In the case of mist, on the other hand, the obscurity is caused by droplets of water, and these will not normally be present in the atmosphere unless the relative humidity is high—perhaps 90 per cent or greater. If the water droplets are sufficiently numerous or large enough to bring the visibility below 1,000 metres, weather people call the resulting murk a 'fog'. Sometimes, to add to the confusion, people refer to light drizzle as a 'mist', a gaffe that will have any thinking meteorologist recoil in horror.

Perhaps it was thoughts like these that Hardy's man was contemplating in his own peculiar way?

THOSE BLUE-REMEMBERED HILLS

24 *April* 2007 ∿

I suppose the ancient maxim 'Blue are the hills that are far away' articulates much the same sentiment as *Bíonn adharca fada ar na buaibh thar lear*. But while the latter is an illusion, a form of self-deception, the former is true in a literal sense, as many of the poets testify. A.E. Housman, for example, asks us:

What are those blue-remembered hills,
What spires, what farms are those?

Milton is a little more obscure when he recalls events 'by slow Meander's margent green and in the violet-embroidered vale', thereby, when you get his drift, contrasting the greenness of the river bank with the bluish hills that form its valley. But meteorologically, Thomas Campbell was nearest the mark when he told us:

'Tis distance lends enchantment to the view,
And robes the mountain in its azure hue.

The key to all this blueness lies in the scattering of light waves by tiny particles in the atmosphere. Waves of any kind are often obstructed by obstacles in their path, just as a boulder interferes with water waves, sending wavelets off in many new directions. In the case of light waves, tiny dust particles in the air, and indeed the very molecules of the air itself, are efficient scatterers of the very shortest wavelengths—those at the blue end of the spectrum. As a consequence, the blue light in any ray of sunshine (which comprises a mixture of all the colours) is diverted in many different directions as it passes through the atmosphere; the remaining colours with longer wavelengths, like red and orange, continue on their journey unaffected.

Now, as you look at, say, a distant mountain, you see, naturally enough, the mountain itself, which typically may be green or brown. But superimposed on this image is another source of light; some of the blue from rays of sunlight passing through the atmosphere is scattered in your direction. It is this scattered blue light which lends the whole scene its characteristic colouring.

The farther away the hills, the more air there is between you and them to scatter light in your direction, and therefore the bluer they will appear. Moreover, in dry, settled summer conditions, when the barometer is high, the number of impurities in the air is higher than usual, allowing a greater degree of scattering and enhancing the bluish tinge. But after a few heavy showers to wash away the dust, the air becomes clear and transparent and the natural colours of the distant hills are relatively unshrouded by any veil of blue.

Also, at the beginning or end of a sunny day, the atmosphere, again because of scattering but in a different way, tends to acquire a pink or orange tinge. It is therefore in the *middle* of the day that the far-off hills are at their very bluest.

BLOODY SHOWERS FROM THE SKY

25 *April* 2007 ⌒

In March 1835, Charles Darwin was exploring in the Andes. 'On several patches of the snow,' he wrote, 'I found the *Protococcus nivalis* or "red snow" so well known from the accounts of Arctic navigators. My attention was called to it by observing the footsteps of the mules stained with a pale red, as if their hoofs had been slightly bloody.'

The history books contain a myriad of references to what people used to think of as being 'showers of blood'. Homer, for example, relates how a 'shower of blood' fell on the heroes of ancient Greece as a harbinger of death. In Italy in 1117 showers of blood caused such a panic that a meeting, not of meteorologists, but of bishops, was held in Milan to consider what their origin might be. And in 1699, according to the records of the French Academy of Sciences, there occurred in several parts of the town of Chatillon-sur-Seine, 100 miles southeast of Paris, 'a kind of rain or reddish liquor, thick and putrid, like a shower of blood. Large drops were seen imprinted against walls, and one wall was even splashed on both sides.'

For the most part, these scarlet visitations were ascribed to divine or satanic interference with the elements. But in the case of the 1699 occurrence, one local commentator at the time had what seems, even today, to be a plausible idea. He noted that since both sides of a wall had been affected, it would 'lead one to believe that this rain was composed of stagnant and muddy waters, carried into the air by a hurricane from neighbouring marshes.' Another commentator got even closer to the truth but hedged his bets:

'What the vulgar call a shower of blood is generally a mere fall of vapours tinted with vermilion or red chalk, but when blood actually does fall, which it would be difficult to deny takes place, it is a miracle due to the will of God.'

Nowadays we recognise several more mundane causes for these happenings. The organisms described by Darwin, or others like *Uredo nivalis* or the algae *Protococcus fluvialis*, often have a reddening effect. There are also a few recorded cases on the Continent of the red deposits being not meteorological but, rather unappealingly, the excrement left by a recent plague of butterflies. By far the most common cause, however, especially when such phenomena occur in Ireland, is that the rain is contaminated by a fine red dust, usually originating in the deserts of North Africa.

The dust is raised in sandstorms and carried northwards by fortuitous winds aloft. It is then washed down to earth by a passing shower, and on the one or two occasions every year when this occurs, we wake to find our cars and other outdoor belongings shrouded in a dappled coat of fine red powder.

WEATHER WHITS FROM WHITBY

27 April 2007 ∿

Today I write to you from Yorkshire, from the fishing town of Whitby, which is dominated by the sombre ruins of Whitby Abbey, set high on a rocky promontory just above the town. Whitby has been mentioned in this column many times, and it is therefore with great pleasure that I see it as it really is.

The date of Easter in these parts, for example, was settled at the Abbey. Back in the seventh century, the Celtic followers of St Columba, firmly established in Scotland, had a certain way of calculating Easter's date, but it differed from the Roman practice followed in the rest of Britain. In AD 664, King Oswy of Northumberland summoned a Council of the two Churches at Whitby Abbey to decide the issue. The Irish lost; the Roman formula, perhaps predictably, was preferred and is still in use today.

The ruined Abbey also figures in Bram Stoker's *Dracula*. The Count, as we know, was Transylvanian, but events require him to visit England. He chooses Whitby as his port of arrival, and conjures up a surreal storm to facilitate his landing: 'The wind fell away entirely during the evening, and then shortly before ten o'clock, without warning, the tempest broke with a rapidity which seemed incredible. White-crested waves beat madly on the level sands, rushed up the shelving cliffs, and swept with their spume the lighthouses rising from the end of either pier at Whitby harbour. Masses of sea fog came drifting inland in ghostly fashion, dank and damp and cold.' Soon 'a schooner with all sails set' is wrecked upon the beach and a big black dog, Count Dracula himself in one of his many guises, jumps out and bounds up towards the Abbey to feed his nasty habit on the unsuspecting Yorkshirefolk.

Whitby's third meteorological connection concerns the appropriately named Dr George Merriweather, the nineteenth-century inventor of the 'Tempest Prognosticator'. This 'instrument' was based on the behaviour of the medicinal leech, which, it seems, relaxes at the bottom of its water bottle in fine calm weather, but half a day or so before a change is due it moves steadily upwards towards the surface.

Merriweather's strange device to harness these meteorological inclinations comprised a central stand surmounted by a bell with twelve hammers. Twelve pint bottles, half-filled with water, were arranged around the base, and each one was fitted with a trapdoor that was in turn connected by a chain to its respective bell. A leech was placed in each bottle, and when it detected a coming storm it would rise to the top of its bottle, and its gymnastics near the trapdoor resulted in a ringing of the bell.

This morning I gazed fondly on Merriweather's brainchild, which is preserved for posterity in Whitby Town Museum.

AN EDINBURGH WEATHER TRAIL

28 April 2007 ∿

Weather Eye and its entourage, currently on tour in northern Britain, has now arrived in Edinburgh, a city rich in meteorological connections. In the mid-nineteenth century, for example, Great King Street contained the house of Thomas Stevenson, lighthouse engineer and father of the novelist Robert Louis. The elder Stevenson is remembered by meteorologists as the designer of the white, louvered box which, even today, houses the thermometers used for measuring air temperature; it is still known as the Stevenson Screen throughout the world.

A few doors away on the same street lived Alexander Buchan, who carried out a detailed statistical study of Scottish weather in the 1850s. He came to the conclusion that certain periods of the year were significantly colder or warmer than they ought to be, and identified six cold periods and three unseasonably warm ones. Meteorologists nowadays treat the 'Buchan's Spells' with scepticism, since more sophisticated analysis over longer periods does not support their regular existence. But even to this day, the relevant periods are eagerly watched by those with an interest in these matters.

But Edinburgh's best-known address is the royal residence Holyrood Palace, and its most famous occupant was the ill-fated Mary Queen of Scots. Born in 1542, Mary succeeded her father, James v of Scotland, when only a week old. She was reared in France and returned to Scotland at the age of nineteen, where, during a short and tragic reign, she was accused of numerous plots to place herself upon the English throne. Obliged to abdicate, she became for eighteen years the prisoner of her cousin, Queen Elizabeth, and was finally beheaded at Fotheringay Castle in Northumberland in February 1587.

Some said her fate could have been foreseen in the events of her arrival in Scotland in August 1561. One of Mary's aides described the happening: 'On the fourth day of the voyage, as the galleys neared the port of Leith, they entered fog. Borne up the Firth of Forth on a fresh east wind, it settled for miles along the shoreline, heavy and impenetrable. All night long the ships floundered outside the harbour, announcing their positions by the beat of drums, until at last the pall lifted slightly with the dawn, and the three vessels crept silently into port to drop anchor.'

The fog was a *haar*, a sea fog peculiar to the eastern coast of Scotland. It often happens in summer that a body of warm, moist air is carried westwards from Europe across the North Sea. As it moves towards Britain, its temperature falls steadily by contact with the colder water; condensation occurs, and a *haar* forms in the easterly breeze playing on the Scottish coast. It was a bad omen for the young Mary, as she took possession of her inhospitable kingdom.

THE FEAST OF CHIMNEY SWEEPS

1 *May* 2007 ∿

Chimneys depend for their effectiveness on the heating of the air by the fire, so that it becomes buoyant and rises up the flue. This effect is augmented by the downward pressure of the cooler, heavier air filling the rest of the room, which forces the warm air up the chimney in much the same way as atmospheric pressure forces mercury up the evacuated glass tube of a mercury barometer. The taller a chimney, the stronger the created draft will be.

Various factors, of course, may interfere with this effective mechanism. The most common is a neighbouring obstruction like a higher chimneystack, a building, or a line of trees which causes turbulence in the vicinity. Once a volume of cold air is forced into the chimney-top by the turbulence, it sinks rapidly downwards to the room below, bringing with it an unpleasant puff of smoke. And problems may also arise from the continual accumulation of particles of soot throughout the length of the chimney flue itself, so that for reasons of efficiency and safety, a chimney must be regularly cleaned.

In pre-Victorian times, until the practice was banned by the Chimney Sweepers' Act of 1840, it was common for chimney sweepers to have small boys in their service who were required to ascend bodily up the flue to scrape away the soot. The value and usefulness of such juvenile assistants depended on their size; ideally they should be very young and poorly fed, criteria which, at the time, it seems, were relatively easy to fulfil.

But chimney sweepers had their good times too. Their traditional holiday in days gone by was today, May Day or 1 May, and on this day they would bedeck themselves in brightly coloured clothes, blacken their faces with their stuff-in-trade, trim their hats and coats with golden ribbons, and go dance and frolic in the city streets.

Chimney sweeps were not the only ones to celebrate. On May Day in ancient times young men would uproot whole trees and set them up in front of the houses of their paramours—no doubt as a none-too-subtle Freudian token of their love. Young girls, meanwhile, would have washed their faces with the May Day morning dew to ensure for themselves a fresh complexion in the years to come:

The fair maid who, the first of May,
Goes to the fields at break of day,
And washes in dew from the hawthorn-tree
Will ever after handsome be.

Eventually, as a way of avoiding the arboreal destruction caused by all this wooing, the authorities introduced the communal notion of the Maypole, around which the local populace would dance after they had crowned as 'Queen of the May' the one adjudged to be the fairest maiden of the neighbourhood.

NARY A DROP DOWN UNDER

2 *May* 2007 ⌒

In Australian Aboriginal mythology, Tiddalick is a giant, ugly frog. From time to time, the Aborigines believed, Tiddalick would swallow all the water in the world, leaving the land baked hard in drought. Pleading for water, all the other animals would gather before the selfish frog, dancing for him and performing tricks for his amusement to persuade him to release the rains. But no matter what their efforts, the frog would bide its time, staring implacably ahead with its eyes and cheeks swollen; finally, when he could hold back no longer, Tiddalick would disgorge the pent-up water, creating massive floods that were even more troublesome at times than the preceding drought.

Tiddalick has been up to his tricks again in recent years throughout Australia. The continent is in the grip of its worst drought on record, with the basin of the Murray and Darling rivers in the southeast of the country being particularly affected. The two rivers normally provide water to an area as big as France and Spain combined, a region which comprises Australia's 'food bowl' and yields 40 per cent or more of the country's agricultural produce. But these rivers are now so low that there will soon be barely enough water for domestic drinking purposes, and Prime Minister John Howard has announced that unless there is significant rainfall over the next two months, a year-long total ban on the use of river water for irrigation purposes will be introduced.

The measure would be a disaster for the country's agricultural sector. Crops would fail; citrus, olive and almond trees would die; and graziers would be forced to sell off sheep and cows.

Thousands of farmers, already laden with debt and in despair after six straight years of drought, would face inevitable ruin.

The causes of the current drought, which began in 2001 but which has been felt most acutely over the past six months, are very complex. The scientific consensus, however, is that climate change has much to do with it, realising predictions that Australia would become hotter and drier in a greenhouse world. Assuming this to be the case, the drought and its potentially catastrophic consequences mark what could be the first disaster caused by climate change to strike a developed nation. And it is not without irony that Australia is one of only two industrialised nations— the US being the other—that refused to ratify the Kyoto Protocol on the grounds that it would damage their economies.

Tiddalick obliged, in his traditional way, a week or so ago. During thunderstorms on 22 April, nearly 100 millimetres of rain fell in just a few hours in parts of Sydney. But the hard ground is unable to absorb such sudden downpours; the result in this case was several flash floods, but no respite for the beleaguered farmers of the Murray-Darling valley.

TRAILING CLOUDS WITH GLORIES

4 May 2007 ~

In religious artistic circles—or is it artistic religious circles?— the terms 'nimbus', 'halo' and 'glory' are virtually synonymous. Indeed the word 'circles' is appropriate in the circumstances, since all three are different names for the bright circular glow which in religious art traditionally surrounds the heads of the most just and righteous. But confuse the three in the company of any meteorologist, and you are likely to be viewed disdainfully down the entire length of an elevated weather-nose.

'Nimbus' is merely the Latin word for a cloud. It is appended, fore or aft, to many of the official names for clouds—like *cumulonimbus* or *nimbostratus*. A halo, too, is a relatively frequent phenomenon, being the large, white, luminous ring sometimes surrounding the Sun or Moon when they are partially obscured by very thin, high cirrus.

But the glory is less common, and for most of us the only opportunity we have of seeing one is when we travel in an aeroplane. It comprises a series of concentric coloured rings arranged target-fashion around the shadow of the aircraft projected onto a layer of cloud below. It is not in any sense *caused* by the plane's shadow; the glory and the shadow are two distinct phenomena, but it so happens that both occur in the same place—at what is called the *anti-solar point*. They are centred at the spot on the cloud which lies directly on an extension of the line joining the Sun and the aeroplane.

The glory is caused by a process called *diffraction*. The mechanism is rather complex, and indeed not entirely undisputed, but in essence what happens is that the water droplets of the cloud interfere with the direct progress of the tiny waves of sunlight, and split the light into its constituent colours—the familiar colours of the spectrum. The light that is diffracted in this way to give the glory is sunlight that has already been reflected inside the droplets of the cloud, in such a way that it returns, more or less, in the direction whence it came.

It is because the light is reflected back along its original path that the colours of the glory can be seen only at that point on the cloud marked by the continuation of an imaginary line drawn from the Sun through the aircraft. And of course it is also at that precise point that you will see the aircraft's shadow.

Glories vary considerably in size, depending on the radius of the water droplets of the clouds in which they have their origin. Indeed as the aircraft proceeds along its track, the glory may vary significantly in size, corresponding to the changing composition of the layer of cloud below. The smaller the water droplets in the cloud, the larger the diameter of the glory.

VOYAGING THROUGH TROUBLED WATERS

7 May 2007 ～

*Q**uis est Scylla?* Or who, for that matter, is Charybdis, who also made a guest appearance on this page on Saturday? And what have either to do with the evolving saga of Election 2007?

Scylla and Charybdis were dreadful sea-monsters of Greek mythology said to have inhabited opposite sides of the Strait of Messina, which separates Sicily from the toe of Italy. Scylla lived in a cave on what is now the Rock of Scylla on the Italian side, and was once a pretty nymph who had inadvertently competed with the enchantress Circe in an *affaire de coeur*. The jealous Circe transformed the nymph into a grotesque and fearsome monster with six long necks equipped with grisly heads, each of which contained three rows of teeth.

Scylla was thus condemned to an existence of misery and loathing, destroying everything that came within her reach. When a ship passed by, each of her heads would seize a sailor from the crew, or as Virgil puts it in the *Aeneid*:

At Scyllam caecis cohibet spelunca latebris
Ora exsertantem et naves in saxa trahentem.

'The cave with its dark hiding place confines Scylla as she stretches forth her jaws to drag ships onto the rocks.'

Charybdis, on the other hand, lived on the opposite shore, a mere arrow-shot away from Scylla. This monster was a giant whirlpool, which first suddenly drank down, and then belched forth, the waters of the strait three times a day, and was thereby potentially catastrophic for any passing vessels.

The challenge for mariners navigating through the strait was to steer a middle course to avoid falling victim to either of these hazards. In Homer's Odyssey, for example, Ulysses is advised by a ghost from the land of the dead to sail closer to Scylla, because Charybdis could destroy his ship. The hero successfully navigates his vessel between Scylla and Charybdis, albeit that Scylla's heads succeed in grabbing a half dozen of his men, devouring them alive.

As it happens, weak residues of Scylla and Charybdis are extant today. The potentially hazardous Rock of Scylla still stands on the Italian side, and the strait's currents do, in fact, present sailors with unusual difficulties. The main current runs from south to north, but a subsidiary current flows in the reverse direction; apparently these currents alternate every six hours or so, causing, *inter alia*, surprising changes in the levels of the water.

More generally in later times, the phrase 'between Scylla and Charybdis' has come to mean being threatened by two dangers in such a way that moving away from one will cause the potential victim to be in even greater danger from the other. *Incidis in Scyllam cupiens vitare Charybdim*, it could have been said of certain events in recent days: 'One falls into Scylla in seeking to avoid Charybdis.'

DEATH IN A COLD SPELL

14 *May* 2007 ∾

'Everyone likes flattery,' said Benjamin Disraeli, 'and when it comes to Royalty you should lay it on with a trowel.' Disraeli's contemporary, Alexander Buchan, may well have entertained some similar thoughts as he contemplated his reply to a letter, which popped through the letterbox of his house on Great King Street, Edinburgh, in May 1879.

The letter was from Queen Victoria. 'Her Majesty has information from someone in America,' it read, 'that from atmospheric and other meteorological observations it would appear that an interval of darkness to be succeeded by a period of intense heat is imminent. She wishes to get information from meteorological observers in this country if they have heard anything of this.' Were these royal premonitions of the greenhouse effect and global warming?

Be that as it may, it is a measure of Alexander Buchan's reputation that he should be consulted on these matters. He was one of the foremost meteorologists of his day, and for more than forty years the presiding genius of the Scottish Meteorological Society. In 1868, for example, ten years after Christopher Buys Ballot had enunciated his famous law about the pattern of winds around low pressure, Buchan refined the notion from his own observations:

'The wind blows neither in a circle round the centre of a depression, nor directly towards the centre, but in a direction between the two. In effect, the direction of the wind at any place makes an angle of 60° to 80° with the line which would be drawn from the place to the centre of the depression.' His assertion agrees closely with today's accepted theory.

But not so his 'Buchan Spells'. Buchan is chiefly remembered for his assertion that at certain regular times of the year the weather is consistently either warmer or colder than the calendar would suggest it ought to be. From a study carried out in the mid-1850s he identified six cold periods and three unseasonably warm ones: the warm spells were 12–15 July, 12–15 August, and 3–14 December; the cold spells were 7–14 February, 11–14 April, 9–14 May, 29 June to 4 July, 6–11 August and 6–13 November. The notion caught the popular imagination, and 'Buchan Spells' were widely accepted for many years as an established feature of our weather.

More sophisticated statistical analysis does not support the existence of such regular occurrences. Indeed, in fairness to Buchan, he himself never claimed that they were a predictively useful guide to the weather, and only suggested that they seemed to apply, at that particular time, in his native part of Scotland. But the good that men do is often interred with their bones. Alexander Buchan died 100 years ago yesterday, on 13 May 1907—in the middle of one of his own alleged cold periods—and is remembered only for the spurious 'spells' that bear his name.

DUNSTAN'S DEALINGS WITH THE DEVIL

19 *May* 2007 ∽

'Foggier yet, and colder. Piercing, searching, biting cold. If the good Saint Dunstan had but nipped the Evil Spirit's nose with a touch of such weather as that, instead of using his familiar weapons, then indeed he would have roared to lusty purpose.'

This Dunstan who makes a guest appearance in Charles Dickens's *A Christmas Carol* needs no introduction to regular readers of this column. He was born near Glastonbury in the first decade of the tenth century, became a monk and rose quickly in the Church. By 945 he was Abbot of Glastonbury, and his wisdom was such that he became chief adviser to the Saxon kings and one of the most powerful men in England in his day. He was consecrated Archbishop of Canterbury in 961.

Now Dunstan, being a talented and clever fellow, became *inter alia* an expert metal-worker, working sometimes as a blacksmith. As the nineteenth-century rhymester Edward G. Flight tells us:

The holy man, when not employed
At prayers or meals, to work enjoyed
With anvil, forge, and sledge.
These he provided in his cell,
With saintly furniture as well;
So chroniclers allege.

But there was another side to the overtly saintly Dunstan. During his time in charge of Glastonbury, where the local apple cider was renowned, he established a brewery in the monastery, and to gain a competitive edge, the story goes, he bartered his soul to the devil in return for an annual spring frost severe enough to blast the apple crop. This, Dunstan reckoned, would stop the local cider-makers in their tracks, and force everyone to drink his Glastonbury ale. Presumably Dunstan managed by some means or other to release himself from his side of the bargain, but the devil has kept his word ever since, frequently giving a regular blast of severe frost in the three days leading up to St Dunstan's Day on 19 May.

Dickens's *Christmas Carol* reference is to another of Dunstan's encounters with the devil. One day, apparently, Old Nick happened by Dunstan's forge and asked Dunstan to repair the shoe on one of his, the devil's, cloven hooves. The future saint, knowing his customer full well, took what he deemed to be the appropriate action in the circumstances:

St Dunstan, as the story goes,
Once pull'd the devil by the nose
With red-hot tongs, which made him roar,
That he was heard three miles or more.

Dunstan released the demon from his grip only on condition that he would never again enter a place where a horseshoe was displayed, which, as we know, the devil has never done even to this very day. Or as the poet puts it:

The horse-shoe now saves keel, and roof,
From visits of this rover's hoof.

A MODEL OF SCIENTIFIC INCONSISTENCY

28 May 2007 ～

'A foolish consistency', according to Ralph Waldo Emerson, 'is the hobgoblin of little minds.' As a young man, however, the Swiss geologist Louis Agassiz was prey to no such lack of flexibility of thought. In perhaps the most celebrated *volte-face* in scientific history, he became an enthusiastic convert to a cause which, at first, he had opposed with vehemence.

Geologists of the early nineteenth century had great difficulty in explaining the existence all over Europe of gigantic boulders which differed in composition from their surroundings, and had obviously been transported to their present location from a distant spot. The generally accepted explanation was that these *erratics*, as they were called, were carried to their destinations by the great currents of water and mud assumed to have accompanied Noah's flood, a theory which conveniently coincided with the word of God, as set forth in detail in the Old Testament.

One Jean de Charpentier, however, a mining engineer who worked in Switzerland, thought differently. In 1834 he became the first to suggest that there might be a tide of ice whose ebb and flow takes half a million years, and which scooped up giant rocks as it advanced only to deposit them in splendid isolation several hundred miles away when it retreated.

Then, enter Louis Agassiz. Agassiz was born 200 years ago today, on 28 May 1807, and as a youthful professor of Natural History at the University of Neuchâtel, the silliness of de Charpentier's ideas so irritated him that he set out to prove them wrong. He carefully measured the movement of stakes he had embedded in the ice on the Aar glacier and, to his astonishment, discovered that the ice moved much faster than he thought it would. He quickly accepted that ice could indeed carry large boulders over long distances and, like Saul when famously Damascus-bound, was instantly converted from staunch opponent to enthusiastic evangelist for this new geological theology.

The campaign against the unbelievers began on 24 July 1837, when Agassiz delivered a very famous lecture to a meeting of the Swiss Society of Natural Sciences, a defence of the new ideas which has become known as 'The Discourse of Neuchâtel'. It began a dispute, one of the most acrimonious in the history of science, that was to rage for more than a quarter of a century, ending with the universal acceptance of the Ice Age theory.

Ultimately, however, the hobgoblins had their way again with Agassiz. He moved to America in the 1840s and became an influential teacher and writer on many aspects of natural history. But in his closing years—paradoxically, in view of his own revolutionary theories—he was one of the most vociferous opponents of Charles Darwin's ideas on evolution, holding that all organisms were immutable and independent of each other. He died in 1873.

ONCE IN A BLUE MOON

30 *May* 2007 ⌁

If you have been looking at the sky of late, you may have noticed that the Moon is waxing. Indeed the full Moon is almost upon us, and will occur around 2am on Friday, 1 June. Moreover, since the lunar cycle is shorter than a normal calendar month, there will be time for another full Moon before the end of June; it will occur on 30 June. A second full Moon occurring in the same month like this is often known as a 'blue Moon', but 'once in a blue Moon' is not as seldom as one might think at first.

The frequency can be roughly calculated. Let us take, for example, a period of 1,000 years. There are 365.242 days each year, so a millennium has 365,242 days, and we also know that each lunation—the period from one new Moon to the next—is 29.5306 days. So our 1,000-year test period contains, dividing one into the other, 12,368 full Moons.

Now we also know that 1,000 years has 12,000 months. It follows that in 1,000 years there must be 368 'blue Moons'—extra Moons to be fitted in as 'seconds' into particular months. And if this occurs 368 times every 1,000 years, it happens, on average, once every 2.7 years. Sure enough, the last time we had two full Moons in the same month was a little under three years ago in July 2004, and before that in November 2001. The next blue Moon will occur in December 2009, and again in August 2012.

It is very unusual indeed, however, to have *two* 'blue Moons' in the one year. This happened in 1999, for example, when after January's double full Moon, February skipped a full Moon altogether, and March again had two. With only 28 days, February is the only month of the year short enough to skip a full Moon; prior to 1999, the last time it happened was 1961, and it will happen again in 2018.

To literal-minded meteorologists, a blue Moon is an occasion when the Moon really does appear to be blue—but it takes a very special set of circumstances to make it happen. The Moon high in the sky is normally white or yellow, and near the horizon, it often appears orange or even red. Very occasionally, however, when many dust particles of a fortuitously specific size are suspended in the air, the full Moon, for a month or two, appears a delicate shade of blue. There are only two recorded instances in recent times: the first was in 1883, after the eruption of Krakatoa, and more recently, in 1950, when the cause was extensive forest fires in Canada. Taken literally, therefore, it would seem, 'once in a blue Moon' means about twice in a hundred years.

THE ELEMENTS AT EPSOM

2 June 2007 ∾

D erby Day, the most famous of the five English Classics, has been held on Epsom Downs every year since 1780, except for two short periods during the wars when it took place at Newmarket, presumably to avoid the odds being skewed by the distracting noise of international conflict. We think of champagne and sunshine, strawberries and cream, parasols, top hats and *My Fair Lady*, and most of the time the elements have acquiesced in this sybaritic image—but not always.

In 1820, for example, the race was run on 18 May and a fierce gale the previous night uprooted tents and booths, leaving a scene of utter devastation for arriving race-goers. Ten years later, in 1830, torrential rain and hail disturbed the scene to the extent that there were no less than thirteen false starts before the horses finally got under way. And concerning Derby Day on 23 May 1867, *The Times* reported: 'The air on the downs from noon was at times piercingly raw and cold, and the holyday makers were suggestively silent by the time they reached Epsom. When the horses got to the post, three-quarters of an hour elapsed before they were despatched on their eventful journey, but the interest attaching to the actual start was considerably marred by the delay and the biting wind, sleet, and snow which swept over the downs; there were at least half a score of false starts.'

But the very worst of Derby Days was that of the 132nd on 31 May 1911, celebrated as the 'Coronation Derby' because King George V was to be crowned a few days later. Much of the month of May that year had been fine, hot and settled, but the weather broke and turned very thundery in the final days. On the afternoon of 31 May the race itself took place in very warm, humid conditions, and was won, appropriately enough, by Sunstar. But shortly afterwards a fierce thunderstorm with 'hailstones as big as walnuts' and lightning at a rate of thirty strokes a minute developed over Epsom Downs, and struck terror into the revellers heading home.

'It was an inferno of water, mud, thunder, lightning and hail,' the *Daily Express* reported. 'Innumerable cars were rendered *hors de combat*, horses plunged with fright, confusing heaps of figures were inextricably jumbled together in narrow roadways, and half-drowned pedestrians, drenched cyclists, terrified women and children, and battalions of men were all helpless against the mighty powers of nature in one of her most savage moods. A boy and a horse were killed at Buckle's Gap, and eight or nine men sheltering on Banstead Downs were struck by lightning, several being killed.' Before the end of Coronation Derby Day, seventeen people and four horses had perished during the storm and in related incidents.

THE WEATHER-WATCHERS OF THE SEA

12 *June* 2007 ∿

W eather observations are the raw material of weather forecasts. Moreover, for countries like Ireland on the edge of Western Europe, one of the difficulties until relatively recently of forecasting the local weather has been the dearth of information from the vast expanse of ocean to the west. Strategically placed weather ships traditionally provided data of this kind, but weather ships are very expensive to operate, and since the beginning of the satellite era there has been a gradual tendency to phase them out. To get the best from satellites, however, it is still desirable to have a good network of accurate instruments at ground level—'ground truth', as meteorologists like to call it; the information from the ground network can then be used to interpret the satellite data more effectively.

In the absence of weather ships, ocean weather buoys have proved useful for this purpose, but even more highly valued are weather reports prepared by human observers on ships at sea as they go about their normal business. The observing ships in question are recruited by national weather services on a voluntary basis, and since the necessary instruments are usually provided by the weather service, which also pays for the transmission of the data, no direct costs are incurred by the ship itself or by its operators. The weather reports obtained in this way are used for climatological purposes, particularly nowadays for monitoring the progress of such climate change as there may be, and also as an input for the computer models used in day-to-day numerical weather prediction.

Back in the 1980s, about 7,000 ships around the world were providing regular weather reports in this way, but the number has declined dramatically since then, decreasing to only 2,500 by the year 2006. The reasons for the decline are manifold. It has occurred partly because of the changing dynamics of maritime operations in recent years, which have involved reduced manning levels, new trading patterns and frequent changes in vessel ownership and countries of registration. Competition among commercial shipping companies has also played a part and, perhaps surprisingly, the increasing frequency of acts of piracy has become an issue; the geographical locations of observing ships, which are of necessity included in their weather reports, often end up on publicly accessible websites, which, it is felt, renders vessels more vulnerable to such attacks.

Naturally enough, efforts are being made to overcome these problems. Solutions being explored as far as the ships themselves are concerned include the increased use of automatic systems and measures to prevent information on geographical locations from entering the public domain. Networks of both moored and freely drifting weather buoys are also being expanded; and in any event, as the quantity and quality of data available from satellites increase, the actual need for surface observations is correspondingly diminished.

THE NILE IN FLOOD

14 *June* 2007 ∼

Ancient Egypt was hostage to the Nile, the entire community depending for its existence on the abundant waters of the stately river. Its level rose and fell with a rhythmic regularity, so that early each summer a surge would move northwards to reach Aswan by the end of June and the delta region by September, before the river subsided gradually to its lowest level by the following April. The rising waters each summer flooded large areas and fed the land with the rich nutrients that made the valley fertile; their failure could, and often did, bring hardship, famine, pestilence and death.

Why this should happen was something of a mystery for centuries. Herodotus of Halicarnassus, for example, confessed himself nonplussed: 'Concerning the nature of this river,' he wrote, 'I was not able to learn anything either from the priests or from anyone besides, though I questioned them very pressingly.' But the wily Egyptians, naturally enough, had their own theories, and presumably saw no reason to share them with this nosy foreigner. They attributed the rise and fall of the Nile to the extra pulling power of the Dog Star, *Sirius*, a prominent feature of the summer sky. And nowadays we have a much more plausible explanation; we assign the seasonal flooding of the Nile to the tropical rains of the equatorial south where the White Nile begins its long journey and, even more importantly, to the melting snows of the Ethiopian uplands where the Blue Nile starts.

But these seasonal floods were never as dependable as might be wished. Moreover, by the end of the nineteenth century, Egyptian agricultural production was inadequate to support the growing population, and the floods, though life-giving, also caused regular damage on the flood plain. To the authorities at the time, it seemed sensible that the Nile floods should be controlled, and so it was that a workforce of 11,000 people raised a great dam just north of the border between Egypt and Sudan. It was over a mile long, and contained a vast reservoir of water for gradual release when needed. This first Aswan Dam was completed in December 1902.

By and large the dam fulfilled its promise. It revolutionised irrigation methods in the region and made possible the reclamation of large areas of arid land for agricultural use; hydroelectric power provided for a large portion of the country's energy needs; and navigation on the Nile was improved by the consistent water flow. But it was sometimes necessary to open the sluices to relieve increasing water pressure, thereby flooding areas it was intended to protect. The ultimate solution was the High Dam at Aswan completed in 1970, an entirely new structure four miles upstream which created Lake Nasser, the third-largest water reservoir in the world.

FRANKLIN'S DANGEROUS EXPERIMENT

15 *June* 2007 ∽

'**H**is greatness sprang more from his practicality than from profundity. In science he was more an Edison than a Newton; in literature more a Twain than a

Shakespeare; in philosophy more a Dr Johnson than a Bishop Berkeley; and in politics more a Burke than a Locke.' And yet despite this pseudo-shallowness described by his biographer Walter Isaacson, it could be argued that Benjamin Franklin is more intimately known today than any of these paradigms.

He is probably best remembered for demonstrating the electrical nature of lightning with a kite, and thereby paving the way for the development of the lightning conductor as a safety device. The kite experiment, which tradition says took place 255 years ago today, on 15 June 1752, was no shot in the dark. Franklin had developed an interest in electricity during the 1740s, and in a diary entry for November 1749 he listed no less than twelve intriguing similarities, as he saw it, between static electricity as generated in the laboratory and lightning as observed in nature.

Franklin had communicated his theories to friends in Europe, together with a methodology for proving his hypothesis. He suggested the use of a 40-foot-high iron rod which, he predicted, 'when such [thunder] clouds are passing low might be electrified and afford sparks'. His instructions were successfully followed at the village of Marly, outside Paris, on 10 May 1752, several weeks before his own experiment, so that by mid-June, although he did not know it, his theories had already been proved correct and his genius was being celebrated throughout Europe.

Franklin himself had intended waiting for the spire on Christ Church, Philadelphia, to be completed before harnessing it for his experiment, but then had the brainwave of using a child's kite. This device was on a silken string, near the lower end of which Franklin had attached a metal key, and when in due course a thunderstorm approached, the fibres on the string began to stand erect. Franklin then placed his knuckle near the key and induced a very considerable spark, confirming his hypothesis that lightning was indeed electrical in nature.

Perhaps the most remarkable thing about Franklin's experiment was the fact that he survived it. Although he was unaware of it at the time, his was a foolhardy exercise, since wet kite strings provide an excellent path to earth for lightning strokes, and in the years that followed, until this became obvious, several people were killed in trying to repeat Franklin's illuminating demonstration. But with the nature of lightning now understood, the use of 'Franklin rods', or lightning conductors, became widespread in the American colonies, their first successful performance being, reputedly, in 1760 when the house of a Mr West of Philadelphia survived unscathed a direct hit by a stroke of lightning.

THE IDIOSYNCRASIES OF THE SUMMER SOLSTICE

21 *June* 2007 ～

This evening at six minutes past seven, Irish time, the Sun will attain the most northerly reach of its annual apparent oscillation north and south of the equator. It will be the summer solstice, the point at which the position of the Earth in its orbit, and the inclination of the planet's axis, combine to allow the noonday Sun to shine directly down upon the Tropic of Cancer at 23° north latitude.

It is traditionally thought of as the longest day of the year. And so it is, in the sense that there are more hours and minutes of daylight today than on any other day, but it does not present us with the 'longest' evening. If you consult sunrise and sunset tables, you will see that the latest sunset does not occur until around 24 June—which, interestingly, is the day on which midsummer is traditionally celebrated—while the earliest sunrise, on the other hand, occurs around 17 or 18 June. But in the midst of this asymmetry, 21 June manages to accrue the most daylight.

To confuse the issue even further, the solstice itself does not always fall on 21 June. You might think it ought to, and indeed, almost by definition, it comes around at exactly yearly intervals, but the fact that our year is not an integral number of days, combined with our odd habit of inserting a leap-day into the annual cycle every now and then to rectify the matter, means that the solstice oscillates backwards and forwards by a day or two according to the calendar.

The term 'solstice' comes from the Latin words *sol* and *sistere*, which in combination might be rendered 'Sun stands still'— something which at first glance appears impossible. Yet anyone willing to take the trouble to observe the rising or the setting sun for a week or two around this time of year can capture the illusion.

The setting point of the sun oscillates up and down the horizon over the twelve-month period. In Ireland, for example, the sun sets almost due southwest around the winter solstice in December, and in the northwest at this time of year. The rate of change of position of the setting-point is greatest at the vernal and autumnal equinoxes; in March, for instance, the sun sets farther and farther north on the horizon with every passing day, and the difference from day to day is greater around that time than, say, a month before or after. By contrast, as the summer solstice approaches, the daily change in the position of the setting-point is very small, until for a few days around 21 June, when it is 'on the turn', it sets at the same place on the horizon every night: in a sense, the setting sun stands still.

| MIDSUMMER FIRE FROLICS

23 June 2007 ~

Tomorrow, 24 June, is the feast of the nativity of St John the Baptist, and traditionally, Midsummer's Day. It was a date of much significance in olden times as one of the four Quarter Days in England, Wales and Ireland, marking the beginning, or end, of the four quarters of the year. The others were Lady Day on 25 March, Michaelmas on 29 September, and Christmas Day, which falls, of course, on 25 December. In bygone times Quarter Days were dreaded or eagerly awaited, depending on your position on the economic ladder; they were the dates on which rents became due, and on which all manner of other financial dealings were transacted.

But the vigil of the feast, on 23 June, is noteworthy mainly for the wealth of superstition which surrounds it, much of it going back to pagan times and presumably related to the summer solstice. Bonfires and torches were of particular importance all over Europe, and indeed still are in many places. Here in Ireland, Humphrey O'Sullivan of Callan, County Kilkenny, was a keen observer of such rituals in the late 1820s, and recorded his impressions in his famous diary, *Cín Lae Amhlaoibh*, translated into English by Tomás de Bhaldraithe.

Amhlaoibh Ó Súilleabháin was originally a Kerryman, having been born in Killarney in 1780 where his father, Donncha, was a hedgeschoolmaster. Father and son moved to County Kilkenny, where at first they taught together, but in due course Humphrey abandoned teaching and became prosperous in business. On 23 June 1827, Amhlaoibh reports: 'The youths and young maidens are dancing around the bonfire,' and the following year: 'Many St

John's torches to be seen on the hills and mountains around about, although there is no good bonfire in Callan. This is unusual.'

Two years later, on St John's Eve 1830, he is more informative: 'There is no bonfire in Callan tonight as a man was killed this time last year at the fire. But from Cnocán an Éithigh at the end of the Green I can see many *sop Seáin* [John's straws] alight. There are flames on every hill and mountain and loud voices to be heard in every green glen around strong fires.'

People would perform various rituals around these bonfires, depending on their priorities in life. In some cultures, it was customary for the young to compete with one another in leaping over the fires to see who could jump the highest over the flames; the winner, it was believed, would be the first of those present to be married, and anyone who jumped clear over the bonfire three times was assured of a happy marriage and a lucky life. For those less ambitious or athletic, merely to walk three times around the bonfire was sufficient to keep disease at bay for a full year.

A DISTINCTION WITH A DIFFERENCE

26 *June* 2007 ∽

Rain, you might think, should be of the genus WYSIWYG. The acronym, you may recall, is What You See Is What You Get; it is computer jargon, and might mean, for example, that what you see on your computer screen will be faithfully committed to paper when you issue the command to 'print'. Similarly, in the case of rain, you might think that if it is wet, comes down from the sky on a very cloudy day, and looks like rain—then surely rain it is?

Not always so! At certain times in the recent past when, as Wordsworth nicely puts it, 'the rain came heavily and fell in floods', you may have observed what you *thought* was rain falling all day long. And yet if you had listened to the weather forecast, rain may never have been mentioned; you may have been told to expect 'occasional showers, with some prolonged and heavy'. But despite the WYSIWYG nonplus, were you to challenge the forecaster, his showers would be defended to the last drop.

The difference between rain and showers, you see, has nothing to do with the duration of the wetting; the distinction is in the clouds. Showers, by definition, fall from individual 'convective' clouds, like *cumulus* or *cumulonimbus*, while rain falls from a 'layer' cloud, like *altostratus*, which is spread more or less uniformly over a large area.

The convective clouds that produce showers have a vertical structure, towering many thousands of feet into the atmosphere; frequently, blue sky and even sunshine may be seen between them. Rain, on the other hand, forms in the relatively flat layers of *altostratus* associated with a weather front, covering the sky like a thick blanket to make the landscape dark and threatening.

As it happens, shower clouds tend to be individual, local phenomena only a mile or two in diameter. So if a cloud is carried along by a fresh breeze, as usually happens, the shower beneath it will not last long in any particular place, but will have moved elsewhere within a quarter of an hour or so. Conversely, rain falling from an extensive area of cloud may take many hours to clear.

Now, a slack area of low pressure of the kind experienced frequently of late is an ideal breeding ground for shower clouds.

Such a pressure pattern, moreover, presents little wind in the upper atmosphere to shift a shower along; it may sit over the one spot for many hours, but the people underneath are being drenched by what is, indisputably, a shower—and not by rain. Conversely again, rain may at times fall from one part of a layer of cloud and not from others, so it may seem to come and go; but despite its intermittent character, it is still rain and not a school of showers.

THE DECLINE AND FALL OF THE ROMAN CLIMATE

27 *June* 2007 ✑

'It was on the night of the 27th of June 1787, between the hours of eleven and twelve, that I wrote the last lines of the last page in a summer-house in my garden. I will not dissemble the first emotions of joy on the recovery of my freedom. But my pride was soon humbled, and a sober melancholy spread over my mind by the idea that I had taken everlasting leave of an old and agreeable companion.'

Thus wrote Edward Gibbon 220 years ago, a short time after he had completed his celebrated *History of the Decline and Fall of the Roman Empire*. The work was published in six volumes and was in its day 'modern to the point of ostentation in its erudition, and innovative to the point of flamboyance in the deployment of its learning'. It was widely acclaimed, but also controversial for its scathing view of Christianity.

Gibbon, by and large, put the decline and fall of Rome down to a gradual loss of moral fibre. 'Cold, poverty and a life of danger and fatigue', he wrote, 'fortify the strength and courage of barbarians. In every age they have oppressed polite and peaceful nations who neglected to counterbalance these natural powers by the resources of military art.' But meteorologists note that the weather was also part of the equation.

They note, for example, that the northward spread of Roman influence coincided with an amelioration of the European climate which had begun about 500 BC. As to why the Romans stopped at Hadrian's Wall and, perhaps, why they never tried to establish their influence across the Irish Sea, Gibbon says: 'The masters of the fairest and most wealthy climates of the globe turned with contempt from the gloomy hills assailed by winter tempests, from lakes concealed in a dark mist, and from cold and lonely heaths over which the forest deer were chased by naked savages.'

But by the end of the fourth century, Europe's climate had begun to deteriorate again and a very severe winter in AD 406 gave an unexpected advantage to Rome's enemies. That year, for the first time for centuries, the Rhine froze over; the Romans still controlled the bridges and crossing points along the river, but ice gave the advancing Vandals unimpeded passage westwards.

Meanwhile in Italy itself, the copious rainfall that had graced the region was diminished greatly. There were poor harvests and periodic famines, and because the previously swiftly flowing rivers of Italy could no longer be maintained, stagnant pools and marshes developed in the river beds and provided ideal conditions for frequent epidemics of malaria and bubonic plague.

And so, in due course, 'the splendid days of Augustus and Trajan were eclipsed by a cloud of ignorance, and the barbarians subverted the laws and palaces of Rome.'

A MONTH OF LUSHNESS

30 *June* 2007 ∿

I t might seem to a casual observer of his verse that William Wordsworth sometimes gets his feet mixed up. Consider, if you will, his little offering which goes like this:

In March, December, and in July,
'Tis all the same with Harry Gill;
The neighbours tell, and tell you truly,
His teeth they chatter, chatter still.

You might think that in the first line the future poet laureate has confused an iambus with a trochee. But no. Until about two centuries ago, the accent in 'July' was on the *first* syllable—so that it rhymed with 'duly'—and also, as the poet rightfully suggests, with 'truly'. When we recall that this is one of Wordsworth's earlier attempts at poetry, written in the 1790s, all is clear.

Be all that as it may, and *pace* Harry Gill, July together with August brings us the warmest period of the whole year. On average, afternoon temperatures exceed 20°C over much of Ireland, and on a few memorable occasions in the past the temperature has soared to 32°C or 33°C. At the other end of the scale, ground frost is very rare in July; it has occurred once or twice, but is so infrequent as to be considered almost a freak occurrence.

Although you might not think so from our experience of late, we are just about to emerge from what, on average, ought to be the driest period of the annual cycle—the months from February to June. But the enhanced power of the July Sun often results in something like a sea breeze effect on a continental scale; there is a tendency for a more westerly drift of wind in the vicinity of Ireland, and a moist air-flow which brings with it more cloud, higher relative humidity and, very often, an increase in rainfall.

But despite, or maybe even because of, the waxing rainfall figures, there is a richness about the sky in July which far surpasses that of any other month. July skies have a unique wealth of detail and variety, perhaps stemming from the accumulated warmth, or maybe associated with the typically high water vapour content of the atmosphere. In any event, one day the sky may be characterised by a deep blue, relieved by a complex pattern of white fleecy cirrus clouds; another by an array of dark threatening *cumulonimbus*, giving clear warning of impending hail or thunder; and on a third the grey monotony of a thoroughly wet day may break just in time to reveal a wild, lurid and fantastic sunset, unmatchable for its complexity and wealth of colour. And a wet July, particularly, causes the vegetation to assume a vividness of tint that far exceeds that which may be displayed at any other time of year.

A BRIEF HISTORY OF CEREOLOGY

2 *July* 2007 ⌒

Hiawatha, wise and thoughtful,
Spake and said to Minnehaha,
To his wife, the Laughing Water:

'You shall bless tonight the corn-fields,
Draw a magic circle round them,
To protect them from destruction …'

To some imaginative souls it might seem at this time of year that Minnehaha may be still at work. For a month or two with the approach of harvest time, perfect circles of flattened stalks appear suddenly in fields of newly ripened corn. Although rare on the Continent, and unknown to my knowledge here in Ireland, corn circles are very prevalent in the south of England, where they have occurred in increasing numbers every year since 1980. Wiltshire and Hampshire are the counties most affected, a large tract of rolling countryside that lies in the triangle formed roughly by Southampton, Basingstoke and Bath.

Within these circles, which may be anything from 3 metres to 30 metres wide, the flattened crop is arranged in a spiral flowing from the centre to the outer edge. The circumference is geometrically perfect, forming an abrupt boundary between the flattened corn within and the vertical stalks of the unaffected crop. Circles sometimes occur singly, but more often nowadays large numbers of them are arranged in complex patterns, seeming to represent, for example, DNA structures, snowflakes, webs or knots, or geometrical patterns associated with the occult or with chaos theory.

Of course, everyone knows where most of them have come from. The craze was allegedly started back in 1978 by one Doug Bower, who claimed to have laid the very first corn circle with a friend as they made their way home one night from their local pub in Wiltshire; they used several planks, some lengths of rope and a ball of string. But the occurrence achieved such wide publicity that other pranksters joined the game, and ever since, the patterns have grown in both numbers and complexity.

But there are those who believe fervently that other influences must be at work as well. They maintain that some of the circles are too complex, or too meaningful, for such a simple explanation. Some like to think of the circles as indicative of a recent visit by a UFO; others like to believe that they may be caused by sudden shifts in the Earth's magnetic field; and you will even find a meteorologist or two interested in this subject.

These last like to be known as 'cereologists', and theorise that the circles may be caused by vortices or whirlwinds. Such vortices, they argue, could not be thermal in origin, because the circles occur in every kind of weather, but may be related to mechanical turbulence which occurs when the wind blows over hills or cliffs; indeed, it has been noticed that, coincidentally or otherwise, many examples occur in the vicinity of rising ground.

Most of their colleagues, however, remain unconvinced.

BEWARE THE COMING OF THE BLIGHT

3 *July* 2007 ~

The humble spud, we are told, is the best package of nutrition in the world, being rich in calories, minerals, vitamins and protein, and virtually free of fat. It is a member of the *Solanaceae* family, which includes tomatoes, aubergines and peppers, and there are about 150 species of it in the wild. Only one of these, however, *Solanum tuberosum*, grows outside the Andes region of South America, although more than 600 different varieties of this are known in Europe.

The only real disadvantage of the potato is its susceptibility to disease, particularly *Phytophtora infestans*, which gives rise to the condition known more commonly as 'blight'. Potato blight is a fungus, often a legacy transmitted by infected tubers from the previous season, and once established, its spread is highly dependent on the weather. For spores to infect plants, three conditions must be fulfilled simultaneously over an extended period: the leaves must be wet and have a film of moisture; the relative humidity must exceed about 90 per cent; and the temperature must be greater than 10°C, ideally within the range 12°C to 20°C. In practice, whenever the latter two conditions are met, it is rarely that the first will not follow automatically.

Light rain and balmy breezes, therefore, provide ideal conditions for infection. Spells of dry weather, on the other hand, are usually free of blight, while at the other extreme, very heavy rain or gales cause potentially infecting spores to be washed harmlessly away. The first signs of infection are water-soaked or pale-greenish spots upon the leaves, which subsequently turn black or brown; affected tubers are soft, dry and very brittle, exude an unpleasant odour, and are, of course, inedible.

Nowadays potato blight can be controlled. The simplest way is to spray the crop with a suitable chemical at frequent and regular intervals from May to September, but this is expensive and adds unnecessarily to the already excessive reservoir of artificial chemicals in the soil. The judicious use of weather information, on the other hand, makes it possible for a farmer to spray *only* when it is necessary to do so, and also—since dry conditions are essential to avoid rain washing the chemicals off the crops—when the operation is likely to be most effective.

The critical parameters, temperature and humidity, are monitored by meteorologists during the blight season, and if simultaneously high values have occurred for some time, and seem set to continue, warnings of the likelihood of blight are issued with radio and television forecasts. These warnings are given together with advice on the suitability of the weather in the coming days for spraying operations. And as it happens, if you have been listening to the weather forecasts over recent days you will be aware conditions are exactly right for potato blight at present.

THE WEATHER IN WONDERLAND

4 July 2007 ∿

Although 4 July is popularly associated with the American Declaration of Independence, it is, in fact, the anniversary of a much more significant event: it was on this date that *Alice's Adventures in Wonderland* were first related to a small but enthralled audience in a rowing boat on the River Isis outside Oxford. The party was led by the Rev. Charles Dodgson, later known better by his *nom de plume* of Lewis Carroll, and included his niece Alice Liddell and her two small sisters.

Now if one were to judge from the internal evidence of the book, one might assume that the *Adventures* had taken place much earlier in the year. Of the March Hare it is said: 'Perhaps, as this is May, it won't be raving mad—at least not so mad as it was in March.' But Dodgson recorded the events of the day in his diary, saying that they rowed from Oxford to Godstow and back, enjoying a warm sunny afternoon on the river, and he places the trip firmly on 4 July 1862, or 145 years ago today.

But was it really warm and sunny? Carroll himself said it was; and Alice reminisced years later about a 'blazing summer afternoon with the heat haze shimmering over the meadows'. But bearing in mind Humpty Dumpty's dictum in a similar context that 'When I use a word it means just what I choose it to mean—neither more nor less', it was highly desirable that independent sources confirmed this vital point. Unfortunately, when literary historians with little else to do looked up the Oxford rainfall records, they discovered that there was a significant amount of rainfall in the twelve hours prior to 2am on 5 July; they concluded, not unreasonably, that the afternoon of the 4th must have been dull and rather wet, and that either the diary entry must be spurious or memories were playing tricks.

This dangerous controversy was finally laid to rest some years ago by Irish meteorologist H.B. Doherty—who at the time must also have had little else to do. Since conventional weather maps were not available for the early 1860s, he consulted archived issues of the London *Times*, and discovered that they contained detailed weather reports from many of the seaports around Britain and Ireland. Using these as one might use modern weather observations, it was possible to reconstruct weather maps for the period in question. The newly constructed charts showed that an active front had passed the Oxford area in the early morning of 4 July, and another moved in from the west late that evening to give the rainfall noted in the records. In between, however, on the afternoon of 4 July, a weak ridge of high pressure provided the idyllic conditions so vividly remembered by the main protagonists.

FOYNES FRETS AND STRUTS ITS HOUR UPON THE STAGE

5 *July* 2007 ～

'I saw the green hills of Ireland and I knew I had hit Europe on the nose.' And further reminiscing about his historic solo 1927 flight across the North Atlantic, Charles Lindbergh continued: 'Ireland is one of the four corners of the world.' He remembered its key geographical position when a few years later he was engaged as consultant to advise Pan American and other airlines about prospective aviation routes to link the Old World with the New.

A transatlantic flight in the 1930s, even between the extremities of both continents, was close to the maximum range of the aircraft of the time. The 'flying boat', however, was by its nature ideally suited to such routes. It enjoyed a wide choice of perfect landing areas, not just on either side of the Atlantic, but at any point *en route* should an emergency occur.

After investigating possible sites in Galway and in Cork, Lindbergh finally settled on the calm land-locked waters of the Shannon Estuary as the perfect stepping-stone for such aircraft plying between Europe and America. Thus it was that an inter-governmental meeting in Ottawa in November 1935 formally agreed the introduction of a scheduled mail and passenger air-service between the two continents, and the little village of Foynes in County Limerick appeared for the first time on aviation maps.

The first attempt at a transatlantic flight from the chosen terminal took place seventy years ago today, on 5 July 1937. The Short S-23 flying boat *Caledonia*, stripped of all non-essential fittings and fitted with additional fuel tanks to extend its nominal range from 760 to 3,000 miles, was under the command of Captain A.S. Wilcockson and took off from Foynes for Botwood, Newfoundland. It completed the 1,900-mile voyage at an average speed of 132mph in a time of slightly over fifteen hours. On the same day, Pan American's Sikorsky *Clipper III* successfully completed a flight in the opposite direction, taking only twelve and a half hours because of the more favourable winds. 'It was a great trip,' Capt. Wilcockson declared on arrival in Newfoundland. 'It shows that a regular transatlantic mail and passenger service is quite feasible.'

And so it was. After many more proving flights over the succeeding two years, it was clear that the journey could be undertaken on a regular basis, despite the challenges of thick cloud, frequent headwinds and occasional airframe icing. The first scheduled commercial passenger flight arrived at Foynes on 9 July 1939, and for a time the little County Limerick village occupied centre stage in transatlantic aviation.

But its hour was brief, if glorious. The last scheduled flight from Foynes took place in October 1945, and with the era of the flying boat almost at an end, operations were transferred across the river to the new Shannon Airport at Rineanna.

| A RAMBLER'S VIEWS ON RAIN

6 *July* 2007 ∽

'Irish rain of the summer and autumn is a kind of damp poem. It is humid fragrance, and it has a way of stealing into your life, which disarms anger. It is a soft, apologetic, modest kind of rain, and even in its wildest moods it gives you the impression that it is treating you as well as it can do under the circumstances.'

William Bulfin, some might say, was taking liberties with the *actualité*, but as he travelled the length and breadth of Ireland on his bicycle, little escaped his beady eye and thoughtful observation. The result was the charming memoir *Rambles in Eirinn*, published one hundred years ago in 1907, a book which in its time had an honoured place on every Irish bookshelf.

Bulfin was an Offaly man, born in the early 1860s at Derrinlough, near Birr. He was a clever chap and well educated, and in 1884 he emigrated to Argentina, where for a time he worked as a *gaucho* on the pampas. William moved in due course to Buenos Aires as a journalist before, in 1901, returning to the family home at Derrinlough, where he died, still relatively young, in 1910.

His bicycle odyssey began 'on the last day of June, and the weather was perfect. The people along the road said it was "shocking warm" and scorching, and "terrible hot, glory be to God". I laid my seven blessings on the Irish sunshine which never blisters, and on the perfumed winds of the Irish summer which are never laden with flame.'

But as we have seen, it rained at times. The rain in Ireland, Bulfin explains, 'does not come heralded by dust and thunder, or accompanied by lightning and roaring tempests, like the rain of the tropics; nor does it wet you to the bones in five minutes. You scarcely know when it begins. It grows on you by degrees. It comes on the scene veiled in soft shadows and hazes, and maybe a silver mist.'

After the arrival of the first tentative drop, which you may think that you have just imagined, 'another comes presently, and you feel it on your cheek. Then a few more come. Then the rest of the family encircle you shyly. They are not cold or heavy or splashy. They fall on you as if they were coming from the eyes of many angels weeping for your sins.

'But they soak you all the same,' allows William Bulfin. 'They fold you in, do those spells of Irish rain, and make of you a limp, sodden, unsightly thing in their soft embraces. But I forgive the rain. It has spoiled many a lovely day for me, but still the memory of it is one that I would not part with for a treasure house of gems and gold.'

THE CLOUD THAT DID FOR PERCY SHELLEY

9 July 2007 ～

Percy Bysshe Shelley knew his clouds. It is doubtful if he would have recognised them by the names we call them now, because that nomenclature did not come into common use until the 1830s, a decade after Shelley's death. But he was a keen observer of the individual cloud-types, and described them in his poems.

Most famously, perhaps, in *Ode to the West Wind* he draws a lyrical pen-picture of the wispy *cirrus* clouds, from the Latin word for 'curl', that progressively invade the sky from the west to bring us

> *... even from the dim verge*
> *Of the horizon to the zenith's height,*
> *The locks of the approaching storm.*

Elsewhere, in *The Cloud*, he describes the giant *cumulonimbus*, which often grows quickly and unexpectedly in the heat of a summer's afternoon and produces showers of heavy hail and thunderstorms:

> *I wield the flail of the lashing hail,*
> *   And whiten the green plains under,*
> *And then again I dissolve it in rain,*
> *   And laugh as I pass in thunder.*

Was it a *cumulonimbus* that did for Percy Shelley 185 years ago yesterday? For the summer of 1822, the poet and his entourage leased the *Casa Magni*, near Spezia on the west coast of the north of Italy. In early May, enthusiastic sailor that he was, Shelley took delivery of a small schooner, which he called the *Ariel*. It had been built at Genoa to the design of a naval friend with speed and style, not safety, in mind.

On 1 July, Shelley sailed the *Ariel* 50 miles down the coast from Spezia to visit his friend, Lord Byron, at Livorno, south of Pisa. A week later, on the afternoon of 8 July, a month before his thirtieth birthday, Shelley and two companions set out on the return voyage. It was a day of dark, still, stifling, louring heat, and at about six-thirty in the evening a violent thunderstorm developed; Shelley and his friends were never seen alive again.

Some say his frail craft, overtaken by the violent squall, simply foundered before the sails were lowered. But years later an Italian seaman confessed on his deathbed to having been aboard a fishing-smack that had rammed the *Ariel* intentionally, with aim of pirating money believed to be on board. And then again, there are those who believe that his death was an assassination, politically motivated.

In any event, several days later the bodies of Shelley and his friends were washed ashore. Shelley was ceremoniously cremated, after the classical Greek fashion, on the beach at Viareggio, his heart was removed and kept by his wife, Mary, and his ashes were interred in the Protestant Cemetery in Rome.

A FATEFUL COMBINATION

13 *July* 2007 ～

There are days when Murphy's Law becomes a *force majeure*, when, as Hamlet puts it, 'sorrows come, not as single spies, but in battalions'. Today, by tradition, is the paradigm of all such days; it is Friday the thirteenth, a day whose unsavoury reputation stems from a combination of the bad luck individually associated with both Friday and the number thirteen.

In the Christian tradition Friday is a day of ill-omen because of its association with the Crucifixion, and also because many undesirable or ill-fated biblical events took place on that day. Eve, for example, allegedly gave the apple to Adam on a Friday morning, and that very afternoon the unfortunate pair were evicted from their Paradise; Noah's Flood, too, we are told, began on Friday.

Even Friday's weather is undependable, as Chaucer tells us in *The Knight's Tale*, recalling that in ancient Rome, Friday was *dies Veneris*, the day dedicated to Venus, goddess of love:

> *Just like a Friday morning truth to tell;*
> *Shining one moment and then raining fast*
> *So changey Venus loves to overcast*
> *The hearts of all her folk; she, like her day*
> *Friday, is changeable—and so are they.*
> *Seldom is Friday like any other day.*

The significance of thirteen in this day's cluster of accumulating woes relates to the history of the number twelve. Twelve was seen as the most 'complete' number. There were, after all, twelve months of the year, twelve gods of Olympus, twelve signs of the zodiac and twelve apostles of Jesus. Thirteen, just one digit beyond twelve, was symbolic of the first departure from completeness, or an initial step towards evil. And in the Christian tradition there is the unlucky significance of the thirteen participants at the Last Supper.

The combination of Friday and thirteen is no less common on the calendar than that of any other day and date. Most years have either one or two Fridays on the thirteenth of a month; this year we have had two—in April and now in July. But some unlucky years have triple trouble with three. This happens in a non-leap year if New Year's Day falls on a Thursday, or in a leap year beginning on a Sunday; it last occurred in 1984, but—luckily?—will not happen again until 2012.

One assumes that it is also leap-year complications that explain the conclusion reached by experts who studied a '400-year repeating cycle' of the calendar; they say that the thirteenth of the month is marginally more likely to fall on Friday than on any other day. Each such cycle contains 4,800 thirteenths; these, apparently, comprise 687 Sundays, 685 Mondays, 685 Tuesdays, 687 Wednesdays, 684 Thursdays and 684 Saturdays. But there are 688 Friday the thirteenths, which seems unfair—or is it just bad luck?

DAYS THAT WERE HOT AND BLACK

19 July 2007 ∽

I n the early eighteenth century, Europe was at the height of the rigours of the Little Ice Age, and the climate of Britain was famously erratic. It was an age of extremes, with harsh winters often alternating paradoxically with long, hot, dry summers which frequently brought drought and hardship. It was during one of the latter, 300 years ago today on 19 July 1707, that there occurred a day which is recalled in the folk-memory of the south of England as 'Hot Tuesday'.

The contemporary scientist William Derham described in his Dawkins-like pamphlet *Physico-Theology* how the day 'called for some time after "the hot Tuesday", was so excessively hot and suffocating, by reason that there was no wind stirring, that divers persons died, or were in great danger of death, in their harvest work. Particularly, one who had formerly been my servant, a healthy, lusty young man, was killed by the heat; and several horses on the road dropped down and died the same day.'

Many other days have been given the same epithet. 'Hot Wednesday', for example, was 13 July 1808, when temperatures of 35°C were experienced in many places in the Home Counties of England, with 37°C being reported in Suffolk. In the United States, meanwhile, there are several references in contemporary writings to the famous 'Hot Sunday' of 18 June 1749. One Dr Edward Holyoke declared this day to have produced 'the greatest heat known in this country at least for the last forty years', while Benjamin Franklin, never one to miss a trick, noted that his thermometer at Philadelphia climbed to an even 100°F that day.

Other weekdays are remembered for being at the other end of the meteorological scale, and are designated 'Black'. 'Black Monday' was Easter Monday 1360; on that day the army of Edward III of England was laying siege to Paris, and it was so dark, windy and bitterly cold that many men and horses died. 'Black Saturday', 4 August 1621, is of Scottish origin; on that day the Edinburgh Parliament had met to reimpose episcopal authority on Scotland, but a violent storm—whether God-sent or not—disrupted the proceedings, and the measure had to be postponed.

And finally, 'Black Sunday' brought one of the worst dust-storms of the tragic dust-bowl years to northern Texas, on 14 April 1935. The reason for the name is obvious from this graphic description written by a schoolboy: 'The storm was like rolling black smoke. We had the lights on all day, and went to school with the headlights on. I saw a woman who thought the world was coming to an end. She dropped on her knees in the middle of the main street in Amarillo, and prayed out loud: "Dear Lord, please give them another chance".'

A SUMMER SOURNESS IN THE AIR

26 *July* 2007 ∿

The kitchen is the very core of any household. Its function has been rigidly defined in legal texts of great antiquity as *Camera necessaria pro usus cookare; cum saucepannis, stewpannis, dressero et stovis inter alia, et omnia pro roastandum, boilandum, friandum et plum-pudding-mixandum.* In such a power-house, naturally enough, chaos ensues if anything goes wrong—like, for example, if the milk turns sour.

Until comparatively recently it was widely held that milk goes sour when there is 'thunder in the air', a phenomenon which was generally assumed to be something to do with electricity. The notion has been forgotten rather than discarded, but in any event, it is a myth. The theory was scotched as long ago as 1913 by two scientists called Duffield and Murray, who showed in a series of experiments that atmospheric electricity, far from turning milk sour, might even act as a preservative.

They drew air through a tube in which were fixed two live electrodes. A high voltage was applied to the electrodes, and the air 'electrified' by the resulting discharge was then allowed to 'bubble' through a flask of milk. The experiments suggested that the acidity—the tendency to sourness—of 'electrified' milk increased much more slowly than that of the ordinary milk over a given period. In fact, if the strength of the electrical discharge was increased beyond a certain point, the level of acidity of the treated milk increased hardly at all, even after several days.

The souring of milk, or the production of *lactic acid*, is caused by the action of bacteria which are called, appropriately enough,

Bacilli acidi lactici. These bacteria are comparatively inactive at temperatures below 7°C, but their multiplication becomes increasingly rapid at progressively higher temperatures, up to somewhere in the region of 38°C. And this explains the association of thunderstorms with sour milk.

The tendency to sourness is caused, not by the proximity of the thunder itself, but by the coincidence of the warm humid conditions which often precede a summer thunderstorm. It is the favourable temperature, rather than electricity in the air, which turns milk sour, an explanation which is strengthened by the relative infrequency of the problem in the case of *winter* thunderstorms.

It is also noteworthy that one of the traditional ways of preserving milk was to place the container of milk in a shallow basin of water, into which a muslin cloth covering the jug was allowed to drape. The water used the muslin like a wick, and its evaporation from the cloth resulted in a drop in temperature. But this method of keeping milk fresh works only when the air is relatively dry, allowing copious evaporation, and would be relatively ineffective in the humid conditions typical of summer thundery weather.

But nowadays, of course, we just put milk in the refrigerator.

THE DECLINE OF THE DEADLY ACID DROPS

28 July 2007 ～

In the early 1980s, before global warming, ozone holes and the El Niño became the popular *bêtes noires*, acid rain had its hour upon the stage as the major villain of the day. The problem was caused by the combustion of fossil fuels and industrial processes which produce sulphur and nitrogen oxides as air-polluting by-products. Chemical reactions in the atmosphere, often brought about by the effect of sunlight, transform these substances into sulphuric and nitric acids, which are then washed to the ground as acid rain.

Acid rain in itself is not directly harmful to humans to any significant extent, but it causes serious environmental damage. It was blamed, for example, for the fact that evergreen trees in many parts of the world were losing needles, turning brown, and dying, and there were dire predictions that entire continents would be entirely woodless. Waters affected by acid rain were often crystal clear, but displayed ugly carpets of green algae and moss which were detrimental to aquatic life. It also gradually depleted soils of calcium and, even more harmfully, reacted chemically in the soil to release aluminium, which caused the local fish to die to from aluminium poisoning. Acid rain also attacks buildings and monuments by causing stonework to deteriorate and crumble.

The problem, of course, has not entirely gone away—but things are better. The issue has been addressed in the United States, Canada and Europe by switching to low-sulphur fuels like natural gas. In addition, in the case of large coal-burning power stations, emissions have been reduced by 'scrubbing' the particulate matter before it emerges from the smoke-stacks. But there is still an acid rain problem at global level emanating from regions with rapidly developing economies, most notably in Asia.

Motor vehicles, too, continue to increase in number worldwide, and produce vast quantities of oxides of nitrogen which convert to nitric acid. Older designs of catalytic converter do not reduce nitrogen oxide emissions effectively, although air quality, in this context at any rate, is expected to improve as older vehicles are gradually replaced.

Emissions in Europe nowadays are governed by various protocols to the 1979 Convention on Long-Range Transboundary Air Pollution. The most notable and recent of these is the 1999 Gothenburg Protocol which requires Ireland, for example, to reduce sulphur emission by around 75 per cent of 1990 values by 2010, and nitrogen oxides by around 50 per cent. As a consequence of this and other measures, European countries have already reported a significant recovery of those forests which were once considered vulnerable to acidic precipitation.

In any event, as the problem has begun to come under control, attention world-wide has shifted to more worrying issues like anomalous floods, storms and heat waves, which it is feared may be associated with climate change and greenhouse warming.

A REVOLUTIONARY WEATHERMAN

30 *July* 2007 ～

Tiflis Observatory was established in 1837 near the centre of the city now more commonly known as Tbilisi, capital of Georgia. It was originally devoted to magnetic measurements, but the observatory's routine was extended during the 1840s to include meteorological observations. By the end of the nineteenth century, after several changes of location, the institution had become, in the words of the Russian meteorological historian A. K. Khrgian, 'the best local observatory in the Empire, from the point of view of equipment, personnel and work'. Its most celebrated staff member, however, was one Iosif Vissarionovitch Djugashvili, better known to history as Joseph Stalin.

Iosif Vissarionovitch's father died when he was young, and his mother, a washerwoman, was determined that her son should be a priest. Young Soso, as he was called, duly enrolled in the Tiflis Theological Seminary at the age of fifteen, where for a time his conduct was exemplary and his academic performance among the best of his contemporaries. But by 1899, Soso's dabblings in subversive movements had been discovered, and he was expelled.

I am indebted to a reader, Roger Timlin, for drawing my attention to Simon Sebag Montefiore's recent biography, *Young Stalin*, which gives details of what happened next. 'Soso needed a job and a home. He became a weatherman. Unlikely as it may sound, the life of a meteorologist at the Tiflis Observatory was a most convenient cover for a young revolutionary. His friend from Gori, Vano Ketshoveli, was already working there when in

October 1899 Stalin arrived to share his small room beneath the Observatory's tower. As a "probationer-observer", he was on duty only three times a week from 6.30am until 10pm, checking temperatures and barometers hourly, in return for twenty roubles a month. On night duty, he worked from 8.30pm to 8.30am, but then he had the whole day off for revolutionary work.'

Stalin's revolutionary activities obliged him in due course to abandon his career in meteorology. 'In March 1901, the secret police, the Okhrana, swooped on all the leaders. They surrounded the weather Observatory to catch Stalin, who was returning by tram. He suddenly noticed through the tram window the studied nonchalance of the plain-clothes secret policemen in position around the Observatory. He stayed on the tram, returning later to reconnoitre, but he could never lie there again.'

Tiflis Physical Observatory, as it was subsequently named, was in due course subsumed into what is now the State Hydro-meteorology Department of Georgia. As Sebag Montefiore describes it: 'The Observatory still stands, though it is as rundown as every institution in Georgia. Stalin's room remains, with a few of his personal possessions and a plaque: "The Great Stalin, leader of the world's proletariat, lived and worked here from 28 December 1899 to 21 March 1901, leading illegal Social Democratic workers' circles."'

| THE MULTICOLOURED MOON

31 *July* 2007 ∾

The Moon is a chameleon; it switches colour through the night to suit the changing circumstances. The present full Moon, if you have noticed, is spectacularly white when it is high in the sky, indicative of a very clear and unpolluted atmosphere—scarcely surprising, since the showery rains of late will have effectively removed a very large proportion of the particulate matter that might otherwise be suspended in the air. This is the Moon's natural colour when viewed without too many atmospheric complications, but when it is rising or setting, very close to the horizon, it is almost orange, perhaps even showing a hint of red.

The white sunlight reflected from the Moon comprises, as we know, a mixture of all the colours of the familiar spectrum, ranging by varying wavelengths from red, through yellow, green and blue, to violet. But if for some reason one or more of the constituent colours is filtered out, the remaining colours add up to something quite different from white. The atmosphere has precisely this effect.

The filtering takes the form of *scattering*. Tiny particles in the atmosphere, and even the molecules of the air itself, interfere with the free passage of the rays of light, diverting them in a different direction from that in which they were originally headed. The short wavelengths of blue light are relatively easily scattered and can be effectively extinguished; the longer wavelengths at the 'red end' of the spectrum are more resilient, and need to follow a longer path through the atmosphere before any noticeable amount of scattering takes place.

When the Moon is high in the sky and shining almost vertically downwards, the path of the light through the atmosphere is relatively short; in this situation only a tiny portion of the light in the very shortest wavelengths of blue and violet is dispersed and the lunar disc appears a yellowy white. The cleaner the atmosphere, the less scattering there will be and the whiter the lunar disc will seem to an observer.

But when we look at the Moon close to the horizon, its light has reached us after passing very obliquely through the atmosphere, having travelled a much greater distance through the air to reach our eyes than when the Moon is directly overhead. This allows much more filtering to take place, and by the time the Moonlight reaches us, its 'bluer' constituents—the violet, blue and green light—may have been almost totally extracted; only the red and yellow wavelengths remain, and these combine to provide the orange colour of the rising Moon.

Then as the night progresses, and the Moon rises higher and higher towards its zenith, the path of its light becomes 'more vertical' and therefore shorter; the colour gradually changes from orange to pale yellow, and ultimately to a very silvery white.

SHEETS, HALYARDS AND THE WIND

3 *August* 2007 ~

If any episode of Victorian fiction caught the popular imagination more than the death of Little Nell in *The Old Curiosity Shop*, it was surely the equally maudlin demise of young Paul Dombey. He, you may remember, was the eponymous sickly son in the Dickens story who very slowly, sad week by even sadder week, succumbed to the rigours of Dr Blimber's rude academy. But *Dombey* has its lighter moments too; one such is when 'Captain Cuttle looking, candle in hand, at Bunsby more attentively, perceived that he was three sheets in the wind, or, in plain words, drunk.'

The phrase 'three sheets in the wind' is commonly believed to be of nautical origin as, indeed, James Joyce implies in *Ulysses*. At one point his protagonists 'set all masts erect, manned the yards, sprang their luff, heaved to, spread three sheets in the wind, put her head between wind and water, weighed anchor, ported her helm, ran up the jolly Roger, let the bullgine run, pushed off in their bumboat and put to sea to recover the main of America'.

It seems the usage can be traced to that strange custom aboard sailing ships of never calling a rope a rope, but esoterically naming it something else entirely. Thus a *halyard*, according to my interpretation of what the experts tell me, holds things— usually sails—vertically; a *sheet* holds things horizontally; and a *line* holds things more or less rigidly in place. Most importantly, the *main sheet* controls the mainsail, and two others—the *windward sheet* and the *leeward sheet*—control something called the headsail.

Now seasoned mariners will no doubt reach immediately for their quills and Quink on reading this shallow, and perhaps erroneous, exposition of the technicalities, but the metaphor is plain enough: if one sheet is loose, a sail flaps in the wind and a ship's progress is unsteady; two sheets 'in the wind', and control becomes extremely difficult; and with 'three sheets in the wind', a ship reels erratically in the manner of a drunken sailor.

Other, however, subscribe to an even more complex derivation. In olden times, if a sheet broke under strain, it was necessary for a sailor to secure the wildly flailing canvas; left to itself, the massive sail would flap itself to pieces. This was a dangerous task; from high up in the rigging, a hapless sailor might find himself crashing to the deck, or swept overboard to almost certain death. The story goes that a volunteer who bravely secured a sheet that had been 'in the wind' was given several generous tots of rum as a reward. A sailor, therefore, who had successfully secured 'three sheets in the wind', and lived to drink his just deserts, was likely to end up happy—but very drunk indeed.

PLOTTING THE PROGRESS OF A PLAGUE

6 August 2007 ∼

Hinc laetis vituli volgo moriuntur in herbis
Et dulcis animas plena ad praesepia reddunt.

You will recognise, of course, Virgil's Georgics III, in which the poet describes a plague affecting livestock in the north of Italy in the first century BC:

And so amidst the springing grass, young cattle die,
And yield their gentle lives at loaded stalls.

His narrative has its echoes in events at present unfolding across the Irish Sea.

Foot-and-mouth disease first appeared in Britain in 1839. Since then it has reappeared only sporadically in these islands, and stringent precautions are taken to avoid its importation. It is a viral illness, affecting, as we know, cattle, pigs and other creatures endowed with cloven hooves. As the name implies, its symptoms include blisters in the mouths and on the feet of the affected animals. Once it is identified, the weather becomes an important factor which may help or hinder the spread of the disease.

The viruses survive in tiny moisture droplets expelled from the respiratory systems of infected animals. These droplets may be carried along by the wind for several kilometres to infect otherwise healthy stock; moreover, the airborne virus thrives in conditions of high relative humidity, but becomes inactive if humidity falls below 60 per cent or thereabouts.

The importance of meteorology lies in the way in which the disease may be controlled. The only effective method is to identify infected areas, prohibit the movement of livestock into or out of them and slaughter all infected animals in the danger zone. The larger the defined areas, the worse the economic consequences, so clearly the aim is to concentrate activity in the smallest areas consistent with isolation of the disease.

Given details of an outbreak, mathematical computer models are used to make useful estimates of the viral concentrations in zones surrounding an infected farm. In a steady breeze, wind-borne vectors will move downwind in a straight line, slowly spreading out by what is called turbulent diffusion. But wind is rarely steady, even for relatively short periods, so a plume of viruses meanders as it drifts. And as weather patterns change over an extended period, so too do the broad features of the wind direction and strength, making the ultimate path of the material very complex indeed.

The existence or otherwise of vertical currents in the atmosphere, which would carry the virus upwards and out of harm's way, are also taken into account. And it is possible to incorporate forecasts of relative humidity into the model, so that the rate of infection for a given concentration of the virus can be assessed.

Using this information, veterinary authorities can then identify specific areas in which to concentrate the eradication effort, avoiding unnecessary and expensive action in zones where the risk is very small.

THE MOST LIKEABLE OF ALL THE BEETLES

13 *August* 2007 ~

'The Creator, if He exists,' opined the Scottish biologist J.B.S. Haldane some fifty years ago, 'has a special preference for beetles.' He was alluding to the fact that there are 400,000 species of beetle on this planet, compared to, for example, only 8,000 different kinds of mammal. Furthermore, although we know that they must have a purpose in the scheme of things, most of us dislike 395,000 or so of these prolific beetle families to the extent that we would just as soon the Almighty hadn't bothered. The 5,000 exceptions are the *Coccinellidae*, species which we fondly recognise as ladybirds.

You can use a ladybird to get a weather forecast. To find the short-term prospects, you sit your ladybird upon your open palm; if it crawls across the hand before dropping limply to the ground, then rain is on the way, but if it ups and flies away in lively fashion, tradition has it that the weather will be fine. The long-range forecast, on the other hand, depends on the number of spots upon the insect's wings; if you have caught a ladybird with less than seven spots, the coming harvest will be good, while one with *more* than seven spots is a harbinger of empty barns. A likely outcome of your search, however, is a ladybird that has exactly seven spots, since that variety is very common, and this, of course, leaves your long-range forecast inconclusive. The spots, apparently, are supposed to remind us of the seven joys and seven sorrows of the Virgin Mary, from whom the insect takes its name.

The proliferation or otherwise of ladybirds in any given summer is both directly and indirectly related to the weather of the previous months. The insects are very sensitive to cold, so a harsh winter depletes their numbers for the following season. On the other hand, and more importantly, a mild winter, like those we have experienced of late, followed by a fertile spring, leads to a healthy crop of the aphids, mostly greenfly, on which the beetles feed, and when well fed, ladybirds, apparently, increase and multiply enthusiastically.

The most memorable year in this respect was that of the glorious summer of 1976. Most of the previous season's ladybirds survived the mild winter, and then a wet spring produced a bumper crop of aphids, and hence in turn of ladybirds. But then a famine followed. By midsummer most of the aphids had been eaten; worse still, the plants on which they thrived had shrivelled in the scorching heat, and by late July every ladybird in Britain and Ireland was starving. In the early days of August that year, billions of them took to the air in search of food, travelled up to 400 miles until they reached the sea, and descended as a well-publicised plague on Britain's eastern beaches.

DARK DEEDS IN DEVON

15 *August* 2007 ~

With much the same logic by which the late Oliver J. Flanagan came to the conclusion that there was no sex in Ireland before television, one might easily conclude that climate change all started round 1980. Certainly, when disastrous and unprecedented floods hit the little village of Lynmouth in Devon fifty-five years ago yesterday, on 14 August 1952, there was no talk whatever of climate change, but there were suspicions that other dark forces might have been at work.

Drought had affected most of the south of England for the second half of July in 1952. The fine spell ended, however, at the beginning of August, bringing a period of changeable weather, with thunderstorms occurring almost daily. Then, on 15 August, a deep Atlantic depression passed directly over Devon, bringing with it continuous, torrential downpours extending over a wide area. During the twenty-four-hour period, 200 millimetres of rainfall were recorded at several spots in the vicinity.

The exceptional volume of water that suddenly arrived that day in Lynmouth was subsequently explained by the unusually large area over which the heavy rain extended. Both the East Lyn and West Lyn rivers, which meet at Lynmouth, rose to quite alarming heights, filled to overflowing with the waters from their Exmoor catchment; peaking at the same time, they were in competition for the limited space available at the confluence of the two rivers. Ultimately, a 12-foot wall of water descended on the hapless town; large boulders and rocks were carried through the streets, destroying houses, roads and bridges, and forming a pile 20 feet in height along the High Street. The search for bodies lasted for several weeks, and thirty-one were found.

But there were strange stories about the Lynmouth flood. In the days preceding the disaster, small aircraft had been seen apparently spraying crops around the area. Then rumours arose that rain-making experiments had been carried out by government scientists at the time, and that therefore the military authorities might have been responsible for the whole disaster. And in fact declassified documents released in 1997 duly confirmed that government aircraft had indeed been 'seeding' clouds around that time.

The technique of cloud-seeding began in the late 1940s when it was found that if powdered carbon dioxide or 'dry ice', or alternatively particles of silver iodide, were dropped into a bank of cloud, they sometimes acted as a catalyst to facilitate a fall of rain. In practice, however, it has always been difficult to assess the efficacy of such a process.

In any event, in the case of the Lynmouth floods the immediate reason for the rain was very clear, in the form of the deep depression hovering over Devon. Cloud-seeding some time previously almost certainly had no significant impact on the unfortunate event—but even today, rumours continue to abound.

EARTHQUAKES, THEN AND NOW

17 August 2007 ~

In bygone times, earthquakes were believed to be caused by the stirrings of some mythical creature buried deep below the surface of the Earth. In Norse mythology, for example, Loki, the former god of strife, caused the chaos from his confinement in a cavern underground. He was chained to a rock with a snake above his head as punishment for all the troubles he had caused the other gods and goddesses. Every now and then a drop of venom fell from the serpent's fangs on Loki's head, and the resulting pain caused him to writhe in agony and cause an earthquake.

In Japan the writhing organism was a catfish or *namazu*. The huge fish produced an earthquake by thrashing its body, but it was restrained in normal times by one of the local gods, who kept a rock pressed down upon its head; sometimes, however, the attention of the custodial deity might wander, and *namazu* gave a massive jerk. And in ancient India it was believed that earthquakes were caused by the occasional stirrings of a giant mole which, for most of the time, slept quietly in its burrow deep below the surface.

Nowadays, of course, we know differently; earthquakes are caused by sudden movements of the Earth's crust. Minor tremors may occur anywhere—indeed they happen in Ireland from time to time—but the more severe ones usually occur near the edges of the major 'tectonic plates' which make up the surface of the planet; the lines where these plates meet—which we know as 'faults'—define the major earthquake zones.

The danger areas comprise a wide zone encompassing the Mediterranean coastlines and continuing eastwards into Asia, the east coast of Asia and the East Indies, and the west coasts of North and South America. In some regions the plates slide past each other smoothly without consequence; in others they do so in a kind of 'stick-slip' motion. Sometimes they may 'stick' for several decades—then slip suddenly by several yards to produce, unexpectedly, an earthquake.

Even without the aid of radio or television, scientists around the world would have been able to pinpoint yesterday's earthquake in Peru with great precision, since distant tremors are detectable with seismographs. These instruments consist, in essence, of a massive block suspended by a spring from a superstructure firmly embedded in the ground. When the Earth shudders, the superstructure shudders with it, but the suspended block—because of its very great inertia—hardly moves at all. The movement of the two parts of the instrument, relative to one other, provides a record of the tremor's progress, and a distant earthquake is detectable as a shock wave affecting a succession of seismographs one after the other, to varying degrees, as it surges right around the world.

WEATHERING THROUGH TROUBLED TIMES

22 *August* 2007 ∾

Valentia Observatory left home in 1892. It was born, as its name implies, on the island of Valentia in County Kerry on the morning of 8 October 1860, when its very first weather report was issued. It was, at first, a very modest establishment occupying a rented house opposite the narrow strait which separates Valentia Island from the rest of Kerry, but as the importance of the station grew and its repertoire of scientific work expanded, it seemed appropriate that it should have a new and much more permanent abode. In 1892 it was relocated to its present site on the adjoining mainland near the town of Cahirciveen, but to everyone's confusion down the years, it retained 'Valentia Observatory' as its name.

And then there came 'the Troubles'. In fact the routine at the Observatory suffered little during the War of Independence. A group of armed men arrived one night in June 1921, and a theodolite and telephone were requisitioned, but telegraphic communications were maintained throughout the period and all normal weather reports were issued without interruption.

But the Civil War was more exciting. Jackie O'Sullivan, in his excellent *Valentia Observatory: A History of the Early Years*, quotes a report written 85 years ago, in August 1922, by the newly appointed Superintendent of the Observatory, C. D. Stewart.

On the 23rd the Irish Free State forces took the town of Cahirciveen after some fighting, most of the actual shooting taking place in the vicinity of the Observatory. The 18h and 21h observations were incidentally rendered extremely unpleasant by the constant crossfire of the two sides. At 11.30pm the occupants of the Observatory were wakened by a party of Free State troops requiring food and shelter. They left about 2am for the town which has since been in their hands.

Apart from the safety of his staff, Stewart's main concern from then on was how to transmit his regular weather reports to London—the *raison d'être* of his establishment. For the rest of 1922 and much of 1923, telegraphic communication was undependable because the wires from Cahirciveen to the Wireless Station at Valentia were regularly cut. A complaint from the meteorological authorities in London, however, provoked a terse response: 'The isolation of this station is no fault of the staff and cannot be remedied by us. Wires calling for the urgent rendering of returns are ludicrous, since to commence with, the wire never reaches us in less than a week, and more frequently takes three weeks.' But by mid-1923, the Observatory routine had settled back to normal, and Stewart was able to remark that it was 'a matter for satisfaction that during the whole of the time during which these considerable difficulties were experienced no observation was missed, and no record was lost from an autographic instrument'.

THE MANY ROLES OF BARTLEMY

24 August 2007 ◦∽

Today is drenched in meteorological significance. First of all, Saint Swithin's forty days expire today, so we might now reasonably hope to obtain parole from whatever sentence he imposed upon us six weeks ago on 15 July. And indeed, give or take a day or two, this would appear to be precisely what has happened, or as the old rhyme puts it:

St Bartlemy's dusty mantle dries
All the tears that Swithin cries.

Bartlemy, you see, is Saint Bartholomew, and you will no doubt have noticed as you said your matins earlier today that this is Saint Bartholomew's feastday.

But there is another less helpful side to Bartlemy. He can be a bit of a Swithin now and then himself, since down in Italy it is believed that if it rains on his feastday, 24 August, further rain is guaranteed on each of the succeeding forty days. Indeed, taken in combination, the two saints can be a troublesome pair since if they were *both* to do their worst, wet weather could be expected for a full eighty days, or the best part of three whole months.

Bartholomew was one of the apostles, and seems to have led a rather uneventful later life, since he is remembered subsequently only for the gruesome manner of his leaving it. In AD 44 he was flayed alive in Armenia, a province *in partibus infidelium* on that isthmus that divides the Black Sea from the Caspian.

In historical terms, Saint Bartholomew's Day has unpleasant

connotations. It was on the eve of this day in 1572 that the Huguenot nobility of France was assembled in Paris for the wedding of Henry of Navarre, the future King Henry IV, to Queen Catherine de Medici's daughter, Margaret. The Queen is reputed to have seen this as an ideal opportunity for a final solution to the Huguenot problem; word of the plot was spread among the populace, and when the bells of the church of Saint Germain l'Auxerrois rang out, the Parisians slaughtered every Huguenot they laid their hands on. Before the Saint Bartholomew's Eve massacre was over, the waters of the river Seine, it is said, ran literally red with blood.

But to get back to weather matters, in mediaeval times today was regarded as the first day of autumn. The accepted passage of the seasons in those days was summed up by the Latin inscription: *Dat Clemens hiemem, dat Petrus ver Cathedratus, aesuat Urbanus et autumnnat Bartholomaeus.* Loosely translated, it was a reminder that that winter lasts from the feast of Saint Clement on 23 November until 21 February, spring from Saint Peter's Day until the feast of Saint Urban, and summer from then until 23 August; autumn, finally, begins today, on the feast of Saint Bartholomew on 24 August.

THE DOWNSIDE OF A GODLY GIFT

27 August 2007 ∾

Poor Prometheus! This minor deity of ancient Greece had so offended the chief god, Zeus, that the latter ordered him to be chained to a rock on the Caucasian mountains. Every day thereafter, an eagle came to eat Prometheus's liver, but since he was immortal, the organ was perpetually regenerated, and so Prometheus was condemned to bear the pain anew on each successive day.

And Prometheus's crime? Apparently, unknown to Zeus, he lit a torch at the golden chariot of the Sun, and sneaking it out of heaven by a subterfuge, he conferred it as a gift on humankind, giving them a singular advantage over other creatures. But as we have seen in modern Greece in recent days, this gift of fire is not without its lethal drawbacks.

There is a global cyclicality to bush and forest fires. Early in the year, in January and February, we hear of them causing havoc in Australia; then, come the northern summer, they are often a serious problem in Spain, the south of France and other parts of southern Europe; and finally, in autumn, it is the turn of the United States and Canada. In a sense, therefore, there is nothing out of the ordinary about the current spate of bush fires raging in many parts of Greece. But the exceptional warmth and dryness experienced in those parts, while we in Ireland endured our dull and damp approximation to a summer, clearly set the scene for a fire season more eventful than the norm.

A bush fire is viable only when the weather at the time, and that of the recent past, have combined to provide suitable conditions. A long, warm, dry spell extracts large quantities of moisture from the vegetation, making it easy to ignite. Once under way, each active fire area provides the necessary heat to evaporate the residual water in adjacent fuel. The current problems in Greece are exacerbated by the seasonal *meltemi*, the brisk, northerly *etesian wind* of the ancient Greeks, which is a regular and usually welcome feature of the Greek summer. Wind in the vicinity of a wildfire, however, greatly facilitates the ignition process by bringing flames and heated columns of air into contact with adjacent trees, and by increasing the available supply of oxygen.

In many of the world's woodlands, unfortunately, human attempts over the decades to control wildfires have ultimately made the problem worse. It has been the practice to try to stamp out all forest fires, on the Shakespearean principle that 'a little fire is quickly trodden out, which, being suffer'd, rivers cannot quench'. But in many places, controlling the smaller fires of yesteryear has allowed the woodlands to become clogged with undergrowth and small trees, which provide, when fire does strike, a vast reservoir of fuel to make the resulting inferno extremely difficult to control effectively.

| THE RIDDLE OF THE TIDES

29 August 2007 ～

The main tidal influence is the Moon. Lunar gravitation produces two 'bulges' on our oceans, one almost directly under the Moon and the other—counterintuitively—on the opposite side of the world; in a sense, this latter bulge can be thought of as being caused by the Earth being drawn away, moon-wards, from its overlying body of water. Half-way between these two mounds of water lie two regions where the sea is a good deal shallower than otherwise it ought to be. As the Earth completes its single revolution beneath the Moon every twenty-four hours or so, the zones of high and low water retain their position relative to the Moon—so two high tides and two low tides should occur at each point on the Earth's oceans every day.

But the Sun complicates the issue. It too causes tides—slightly less than half the size of the lunar tide in magnitude. Sometimes the two tidal effects work against each other, and at other times they co-operate—notably at the new and full Moons, when the Earth, the Moon and the Sun are more or less in line.

When these astronomical parameters are fed into a computer, it is easy to calculate what is known as the Equilibrium Tide—a theoretical tide appropriate to a landless world, uniformly covered in deep water. But the observed tides differ greatly from this Equilibrium Tide. Each ocean is, in a sense, an isolated bowl of water; and because of its dimensions, depth, and general shape, and the interfering walls of land around its edges, it responds more readily to certain tide-generating forces than it does to others.

In a similar way, every inlet of the sea has its individual character, determined by its shape, depth and orientation.

Sometimes it happens that the physical characteristics of a bay cause one of the components to *resonate*—to amplify out of proportion to the others and to dominate—giving a particular inlet a tidal pattern which is distinctly different from any others in the neighbourhood.

In practice, to prepare predictive tide tables for a particular port, it is necessary first to install a tide gauge there, and to record the observed pattern of the rise and fall of water over an extended period. Careful examination of such a record allows it to be recognised as the rather complex combination of a large number of independent simple rhythmic undulations, each one corresponding to the known characteristics of one of the individual astronomical components. When the relative strength of each astronomical component at a particular spot is empirically identified in this way, it can then be given the correct emphasis when constructing the mathematical equation from which a computer can calculate the tide at that place for any future time.

THE RIDDLE OF THE TIDES 2

30 *August* 2007 ～

The tides are the pulses of the world's oceans. They are most obvious at the ocean rim, where their rhythmic ebb and flow brings about that periodic immersion and exposure of the intertidal zone which has such profound effects upon indigenous ecosystems. And the tidal currents, accompanying the rise and fall of the water level, are by far the strongest currents in the coastal ocean; even today's large and powerful ships prefer to leave a port with an outgoing tide, and to enter on an incoming one, just as the sailors of antiquity did in times long past.

The Chinese regarded seawater as the blood of the living Earth, and the tides as the beating of its pulse. The Arabs, on the other hand, correctly focussed on the Moon; they supposed the Moon's rays to be reflected from the rocks at the bottom of the sea, thus heating and expanding the water, which subsequently rolled in waves in the direction of the shore. And in more modern times, the seventeenth-century French philosopher Descartes leaned in the direction of Aristotle's theory of the elements. Descartes was of the view that space was full of an invisible substance known as ether, and that as the Moon travelled on its journey around the Earth it compressed this ether in a way which transmitted pressure to the sea—and caused the tides.

But then in 1642, Isaac Newton was born, and it was he who was to put an end to tidal speculation. He applied his formulation of the law of gravitational attraction to explain the influence of the Moon and Sun, and the riddle of the tides was solved.

As regards the influence of the tides on weather, the only advice of any consequence is that given to the farmer in a little rhyme which emphasises the perceived beneficial effects of rising water:

If it raineth at tide's flow
You may safely go and mow,
But if it raineth at the ebb
Then, if you like, go home to bed.

Meteorologists, as always, are sceptical of such relationships. If the ebb and flow of tides has any effect upon our weather, they argue, it must be very minor and extremely localised. It is possible, for example, that tides may have some influence on the timing of the onset or disappearance of coastal fog. As the sea retreats with the ebbing tide, it often leaves exposed a large area of sandy foreshore whose thermal characteristics are quite different from the water surface that was there before; this might make fog more likely at certain times in the tidal cycle than at others. Apart from this almost trivial example, however, there is no convincing evidence that the tides affect our weather in any way at all.

THE SYMPTOMS OF SEPTEMBER

3 September 2007 〜

Meteorologists base their concept of the changing seasons on the average temperature experienced at these latitudes. They designate the three coldest calendar months as 'winter' and the three warmest ones as 'summer', so almost by default, September, October and November turn out to be the autumn months. By this reckoning autumn 2007 began on Friday.

This autumn follows what by any reckoning has been one of the most dismal summers for many, many years, and the official statistics confirm the conclusion reached by even the most casual observers. Virtually everywhere in the country was wetter than it ought to have been, and many parts of the east had well over twice their usual summer rainfall quota. Only in the last days of August did we experience any semblance of what we might think of as normal summer weather.

The continuation of this late summer fine spell, albeit with some interruptions, has given autumn and September a very gentle *entrée*. But September is a month of slow transition, and as we return to school and work, and check the central heating, it will become increasingly obvious that the year is on the turn, and that the slow and sad decline into the aches and pains of winter has begun.

When September weather is quiet, the lengthening nights make it an ideal month for the dews and mists we associate with early autumn; the nights are cool and ground frost becomes a common occurrence as the month matures. But September can be ambivalent and indecisive. While there may be fine spells, as at present, it can also be a windy, blustery month, characterised by a regular procession of depressions moving eastwards across the Atlantic and passing close to Donegal and Mayo. Occasionally, one of these may harbour the atmospheric remnants of some almost forgotten transatlantic hurricane, as, for example, did the storm we recall as Hurricane Debbie on 16 September 1961. Many of the records for extreme wind-speeds established on that day still stand.

Which brings us to another interesting feature of our recent summer: its dearth of hurricanes. Until a day or two ago there had been only five named storms in the North Atlantic, and of these only Dean matured to full hurricane status; it caused considerable havoc in Jamaica before losing most of its strength crossing the Yucatán peninsula. Erin fizzled out without doing any harm to anyone, but at the time of writing a vigorous Hurricane Felix is following a track very similar to Dean's.

As recently as 9 August, the American authorities predicted a 'very high likelihood of an above-normal 2007 Atlantic hurricane season', with somewhere between thirteen and sixteen named storms expected to occur. No revised opinion has appeared, so clearly, if they are to be right, September must be the month to watch for hurricanes.

AUGUSTE COMTE'S CURIOUS CALENDAR

5 September 2007 ∾

Mankind has had a plethora of calendars. The most enduring of them in this part of the world have been the Julian Calendar, in use since 46 BC, and the Gregorian Calendar, which replaced it several hundred years ago. The French Republicans introduced what they thought was a more rational system of reckoning the date in 1792; and perhaps the most bizarre of all to retrospective eyes was the calendar proposed by Auguste Comte in the middle of the nineteenth century.

Comte, a Frenchman, was born on the first of *Pluviose* in the sixth Year of the Republic, or 19 January 1798, as you or I might call it. In his younger years he was something of a rascal. He went on, however, to be widely respected as a mathematician, and it was only when he began to dabble in philosophy in middle life that his ideas became a bit eccentric. He went on to propose the formal worship of a Being called 'Humanity'—a collective concept epitomising those who had seemed to be devoted to the well-being and the progress of the human race. He composed a list of those deemed worthy of the kind of reverence given to the saints and prophets of more orthodox religions, and his calendar was intended as a kind of mnemonic for this exercise.

'All the relation to the moon being set aside,' said Comte, 'and the month becoming as subjective as the week, we soon come to see that it is necessary to make the month invariably four weeks exactly, which leads to the division of the year into thirteen months. The complementary day with which, on this system, each year ends, will have no weekly or monthly designation, any more than will the additional day which follows it in leap years.'

And so it was. Comte's Positivist Calendar, unveiled in 1849, had thirteen months of four weeks each, and was arbitrarily designated to have started in 1789, making this year, 2007, the year 219 by his reckoning. The months were dedicated in turn to Moses, Homer, Aristotle, Archimedes, Caesar, Paul of Tarsus, Charlemagne, Dante, Gutenberg, Shakespeare, Descartes, Frederick the Great of Prussia and Voltaire. Fifty-two more names were associated with the weeks, and 365 others with the individual days. The entire litany of 424 names was a compendium of gentlemanly excellence in philosophy, science, literature, statesmanship and war, while to add a token sense of gender balance, the extra day in a leap year was consecrated to the worthy and collective notion of 'Good Women'.

Unfortunately—or perhaps fortunately for those of us who find it difficult to remember dates—Comte's Positivist Calendar never quite caught on. Auguste Comte's death is remembered in the orthodox way as having occurred 150 years ago today, on 5 September 1857.

THE GREAT DESERTS OF THE SEA

6 *September* 2007 ∿

Jonathan Swift noted that certain parts of the world seem rather empty:

So geographers in Afric maps,
With savage pictures fill their gaps,
And o'er uninhabitable downs
Place elephants for want of towns.

But the cartographers cannot be blamed for these anomalies. It just so happens that certain regions of the world are prone to deserts, which by their very nature are bereft of towns, people and other signs of habitation. These great arid regions of the world are inhospitable because they are dusty, hot and very, very dry— broad oceans of trackless, burning, virtually lifeless sand. But there are other kinds of desert—deserts on which next to no rain falls, like those above, but which are nonetheless, quite paradoxically, awash with limitless supplies of water. There are the deserts of the sea.

Over each hemisphere of the Earth, between latitudes 25° and 45° or thereabouts, are two vast semi-permanent belts of high pressure girdling the globe. They are separated from each other by the rainy equatorial zone of low pressure, the doldrums, and they move northwards and southwards with the seasons, being about 5° of latitude closer to the equator in their corresponding hemisphere in winter than they are in summer. It is the periodic, and it must be said anomalous, extension of one of these, the Azores High, that gives us here in Ireland a brief taste of fine weather from time to time like that which we enjoy at present.

These zones of high pressure are, in general, areas of light winds and gentle subsidence. The subsiding air is warmed by compression as it sinks, resulting in a very low humidity, so rain-bearing clouds are comparatively rare, the weather normally is fine and sunny, and land areas beneath experience a very arid climate.

It is not surprising, therefore, that most of the world's great deserts lie in the latitudes affected by these semi-permanent highs.

They include in the northern hemisphere the Sahara of North Africa and its extensions into Saudi Arabia and the Middle East, and the deserts of the southwest United States and adjacent parts of Mexico; in the southern hemisphere are the Atacama of Chile, the Kalahari of southwestern Africa, and vast tracts of inland and western Australia.

But this aridity does not confine itself to land. There are also vast expanses of ocean under these high-pressure zones where rain of any consequence is almost non-existent. Decades may go by with hardly any precipitation in that part of the South Pacific stretching westwards from Peru and northern Chile, a corresponding area to the west of Angola in southern Africa, and in the northern hemisphere a region stretching westwards from the Sahara out into the Atlantic.

These are the great deserts of the sea.

| THE SYDNEY BRICKFIELDER

7 September 2007 ~

To judge by *Neighbours*, Australia seems like a pleasant little spot. Most of us, however, do not know a great deal about it, and our impressions might be summed up in the words of the Duchess of Berwick in *Lady Windermere's Fan*: 'It must be so pretty with all the dear little kangaroos flying about.' For the meteorologist, however, crossing the equator can be a traumatic experience, presenting a world where all the familiar concepts are turned upon their heads.

One can cope with the fact the seasons are reversed; Christmas comes in the middle of summer, and what passes for winter in Australia occurs in July and August, but the fact that all their fronts and depressions are upside-down and back to front takes getting used to.

The northern visitor may also be confused by funny names. Some old-timers call their hurricanes *willy-willies*, and experience strange phenomena like 'Cock-eye bob', a sudden squall like a tornado which occurs off the northwest coast. The fresh and gusty 'Fremantle Doctor', on the other hand, brings a welcome relief to the neighbourhood of Perth on very hot antipodean summer days.

But perhaps the most evocative Australian wind phenomenon is the *southerly buster*, or 'burster' as the more refined among us like to call it. The buster is a surge of cool air that moves rapidly northwards from the polar zones and then along the southeast coast towards Sydney in the summer months. It is accompanied by squalls and a sudden, rapid drop in temperature, and the gale force winds accelerate and intensify as they move inland towards the highlands of New South Wales.

Sydney has about thirty southerly busters every year. The sudden onset, and the characteristic roll-cloud that accompanies a buster, was described graphically more than 100 years ago by an Australian meteorologist called Henry Hunt: 'Afar off this cloud is sharply defined, dark on the edges with lighter shades in towards the centre. The roll is from thirty to sixty miles in length, and as it approaches, the wind which has been blowing from the north drops suddenly. Immediately under the roll, light clouds rush forward with great velocity only to be thrown back over the top as they reach the front; the wind vane on the Time Ball tower flies to the south, and the wind reaches us on the ground a moment later, and in a few moments is blowing with the full force of a gale.'

In the early days of the Australian settlement, a southerly buster approaching Sydney was always heralded by a cloud of reddish dust that emanated from the extensive brick-fields in the city's suburbs. The brick-fields in that part of Sydney are long gone, but the local name for the phenomenon they engendered is still occasionally heard: the *brickfielder*.

THE PUNY AND ELUSIVE MOONS OF MARS

8 September 2007

Long before they were actually discovered and looked at through a telescope, it was confidently believed that the planet Mars must have two moons. The notion is believed to have originated with Johannes Kepler, who based his somewhat tenuous inference on the fact that, in his time, Earth was known to have one Moon, Jupiter four moons, and Saturn five, and so the rhythmic order of the universe required that Mars have two. In the early 1600s he wrote a letter of congratulation to his colleague Galileo, who had just discovered the moons of Jupiter, and remarked in passing: 'I am so far from disbelieving in the discovery of the four circumjovial planets, that I long for a telescope to anticipate you, if possible, in discovering two around Mars.'

The notion persisted through the centuries. The French philosopher Fontenelle, for example, mentions the possibility of two Martian satellites in *Entretiens sur la pluralité des mondes* in

1686. Then Voltaire, writing in 1752, declared that voyagers to Mars 'would see the two moons which belong to it, and which have escaped the searches of our astronomers'. Even our own Jonathan Swift in *Gulliver's Travels* describes how the inhabitants of the floating island of Laputa 'have likewise discovered two lesser stars, or satellites, which revolve about Mars'.

But even if they were suspected to exist, the moons of Mars proved decidedly elusive. The planet's orbit relative to Earth is such that conditions favourable for a sighting occur only every fifteen to seventeen years. On one such occasion, in the year 1862, a systematic search was carried out, but yielded nothing.

Fifteen years later, however, persistence was rewarded. A new 26-inch reflecting telescope installed at the Washington Observatory under the stewardship of a young American called Asaph Hall greatly improved the chances of success, and at 2.30am on the morning of 12 August 1877, Hall spotted the first Martian satellite. He observed the second one less than a week later on 17 August, and named the pair Phobos and Deimos, or 'Fear' and 'Terror', after the mythological sons of Mars, the Roman god of war.

The Martian moons, however, are but puny, tiny things. Phobos, the larger of the two, is around seventeen miles in diameter, and Deimos is only nine miles across—not much bigger, for example, than Valentia Island. During the Martian night, Deimos would appear as nothing more than a rather large, dim star. And Phobos is very close to its parent planet, being no farther from the surface than Dublin is from Baghdad; it therefore takes less than eight hours to complete an orbit, and to an observer on Mars it would appear to rise in the west and skim rapidly across the sky to set in the east a little over four hours later.

SURVIVAL IN A VEGETABLE WORLD

10 *September* 2007 ～

Before the last Ice Age, the region we know now as Britain and Ireland had much the same variety of trees and other vegetation as it does today. But then the Arctic ice advanced over the area, and modern France and Germany were reduced to a state of permafrost and few, if any, trees withstood the southward onslaught of the ice sheet; the familiar species survived during this Ice Age only in little pockets of what we would now call a 'temperate' climate on the Mediterranean peninsulas—Italy, the Balkans and Iberia. Then, from about 10,000 years ago, as the ice retreated, the trees and plants began the slow process of repopulation, providing once again a dense covering of forest.

This prehistoric retreat and advance of the vegetation suited to a certain clime is an example of how Nature allows species to evolve and die away, to spread luxuriantly or to scurry off to more amenable surroundings when conditions locally are not entirely to their taste. Individual plants, on the other hand, have developed ingenious techniques to maximise their opportunities while staying where they are.

On a global scale, the stature and complexity of vegetation, in general, increases with the level of precipitation. Where water is plentiful, the vegetation is dense and lush; where moisture is in short supply, plant-life is sparse and stunted. In equatorial regions the abundance of rain allows the tallest and most luxuriant of trees to thrive in the evergreen rainforests, but this abundant rain is not without its disadvantages.

Leaves, as we know, are the solar panels of a plant; they soak up sunlight and, through a process known as photosynthesis, use it to create the material for growth. In the dark rainforests of the tropics, sunlight is in short supply for many of the lower plants, and they must take full advantage of every speck that filters through the foliage above. A film of rain can interfere with this objective: it reflects a large proportion of any light available.

One solution to this problem is for the plant to evolve leaves of such a shape that the water drains away—leaves that are often equipped at the ends with downward-pointing spikes that act like eaves. Other plants secrete an oily coat of resin to repel the water; the rain either runs off the leaf immediately, or else accumulates in separated drops, rather than covering the surface in a film. And another trick is for the leaf to cover itself with tiny hairs; they act as umbrellas, and prevent the drops from making contact with the surface. Indeed, in these circumstances the drops can then act as little lenses which, by focussing the sunlight, dramatically increase—in certain spots—the light available to the plant for photosynthesis.

NEW YEAR'S EVE FOR SOME

12 *September* 2007 ∽

Milton, in *Paradise Lost*, tells us exactly how the world began, and describes Creation in terms of one vast twirl of the Almighty's compass:

One foot He centre'd, and the other turn'd
Round through the vast profunditie obscure,
And said, thus far extend, thus far thy bounds;
This be thy just Circumference, O World.

Needless to say, there are many differences of opinion, *inter alia*, as to when this great event took place. The Hebrew tradition, however, follows the Talmud, the body of scripture on which Jewish civil and religious laws and moral codes are based, and it is deducible from this, it seems, that some 4,000 years separated the Creation from the destruction of the Temple at Jerusalem in AD 69.

Further refinements place the Creation firmly on 7 October 3761 BC, as reckoned by the Christian Calendar, and it was on this important premise that Rabbi Hillel II, back in what we now call the Gregorian *Anno Domini* year of 358, based the long-enduring Jewish Calendar. Today is the last day of the year 5767 of the epoch *Anno Mundi*, or 'Year of the World'.

The *Anno Mundi* system is described as 'luni-solar', in that the length of each individual year varies depending on the motions of both Sun and Moon. Most Jewish years consist of 12 months of alternately 30 and 29 days each. Since this, however, adds up to only 354 days altogether, it is necessary every now and then to add an extra month, to ensure that the calendar does not diverge too much from the year as indicated by the Sun and seasons. The result is an occasionally inserted 'embolismic' year—one that has 13 months and comprises 383 days.

Rosh Hashanah, the Jewish New Year, occurs annually around this time, but its precise date depends, not only on these leap year and leap month complications, but also on the first appearance of the New Moon around the time in question. And there are further socio-religious refinements if the beginning of a year is set to fall on a Sunday, a Wednesday or a Friday; in such cases Rosh

Hashanah is postponed until the following day. In ancient times, official witnesses were sent to watch for the arrival of the New Moon, and only on the basis of a formal report of their vigil could the bonfires be lit on the hilltops to tell the people that the New Year had begun.

Nowadays, however, the date of the New Moon's first appearance can be calculated astronomically with great accuracy. Tomorrow, then, 13 September, will be the first of Tishri, the initial month on the Jewish Calendar and the first day of AM 5768.

HIGHLIGHTS FROM THE HOTSPOTS

13 *September* 2007 ∽

A little place called Al'Aziziyah in Libya achieved its place in history eighty-five years ago today. There, on 13 September 1922, the mercury in the official thermometer rose to 58°C, a clear world record, both before and since, and a figure yet to be equalled despite our current fears, and indeed experience, of global warming.

Despite this notoriety, Al'Aziziyah is not reckoned as the hottest place on Earth. This latter accolade goes to Death Valley in southern California, a deep gorge 145 miles long, surrounded by tall mountains that soar to 10,000 feet or more above the valley floor. The reputation of the aptly named crevasse as a hell on Earth was sealed in July and August 1917 when the temperature in the valley—measured in the shade, of course—exceeded 50°C for 43 consecutive days. The place on Earth with the highest *average* temperature, on the other hand, is reckoned to be the Danakil Depression in Ethiopia.

Now all these places have a number of features in common. Clearly, low latitude plays a part, since at locations above about 45° north or south latitude, the Sun's rays approach the surface of the Earth at too oblique an angle, even in high summer, for them to result in extremely high temperatures; the energy is distributed over a wider area than would be the case if the radiation were beaming down nearer the vertical. Another important ingredient is lack of wind; wind causes turbulence, which in general prevents temperatures from rising as high, or indeed falling as low after dark, as they would in its absence. And the desert sands are important, since the air warms more rapidly over such a surface than it would, for example, over a grassy meadow.

But the final ingredient is that the really hot places on the Earth's surface lie in depressions, where gently moving air sliding down towards the valley floor is compressed by increasing barometric pressure, and therefore becomes increasingly warmer and drier. Al'Aziziyah, for example, is not, as you might suspect, at the very heart of the Sahara Desert, but a depression a mere 100 miles or so inland from the Mediterranean coast near Tripoli; Death Valley, too, is also remarkable for being the only part of North America where the land is below sea level; and the Danakil Depression is not near the centre of the African landmass but a mere 100 miles inland from the Red Sea.

Temperatures like these, as we know, do not occur in Europe, although the highest official temperature ever registered on the continent—50°C on 4 August 1881, in Seville, Spain—came close enough. But the highest experienced on the island of Ireland since instrumental records began about 150 years ago was the 33.3°C recorded at Kilkenny Castle on 26 June 1887.

TODAY, IN ONE SENSE, THE SHORTEST DAY

17 September 2007 ∾

The evenings, as they say, are 'drawing in'. After the summer solstice, the time of sunset advances a little every day. The daily change is very gradual at first, but as the months progress, the evenings shorten at an ever-increasing rate, until by around this time of year, sunset each night becomes earlier by a full quarter of an hour per week. Later we will see the rate of advance begin to ease again, and the change from week to week will again be very gentle as we approach the winter solstice in December. What it means, in fact, is that during these, the second and third weeks of September, the evenings shorten more rapidly than at any other time throughout the autumn.

But there is another sense in which September days are short. The twenty-four-hour day by which our lives are regulated is based on the average length of time between two consecutive transits of the sun across the same meridian—if you like, the average length of time from noon to noon. But this interval is not exactly constant: the length of the 'solar day', as it is called, varies rhythmically throughout the year—one of the reasons being the fact that the Earth moves slightly faster in its orbit at certain times, when it is closer to the sun, than it does at others.

The solar day is at its longest around Christmas—at about 24 hours and 30 seconds. It gradually shortens until late March, when it is only 23 hours 59 minutes and 41 seconds long, and then the solar day lengthens to reach 24 hours and 13 seconds by mid-June. During the following three months the day shortens again, and the very shortest day of the year, a mere

23 hours and 38 seconds in length, occurs on or about today, 17 September.

And there are fluctuations in the length of day over longer periods as well, as the Earth, for various reasons, increases or slows down its rate of spin. Careful measurements have shown, for example, that in the last 100 years or so the average day was at its shortest in the 1860s, when it was eight-thousandths of a second less than in the early 1900s; from 1910 until about 1930 the Earth accelerated, shortening the average day, and then slowed down again to provide the longest average days in recent times during the early 1970s.

And to confuse the issue even further, the days are growing longer all the time anyway as the Earth grinds slowly, very slowly, to a halt because of the frictional effect of the lunar tides. Five hundred million years ago, each day was only a little more than twenty hours in length, and there were 430 days in every year.

LIQUID UNDERGROUND ARTISTRY

18 September 2007 ∿

Beneath the fertile undulating fields of Munster and of Meath, and the rich marshy bogland of the central plain, there lies an underlay of solid rock. In places it breaks through, and there is little else: this happens, for example, on the Burren in County Clare, whose nakedness moved one Cromwellian general, mindful of his trade, to remark that here was not 'water enough to drown a man, nor trees to hang him from, nor even enough earth in which to bury him'.

The rock is of three basic kinds. *Igneous* rock, as its name suggests, has its origins in fire; the granite of the Wicklow mountains and the basalt of Antrim and the Giant's Causeway originated from hot molten material that, aeons ago, surged up from seething cauldrons in the bowels of the earth. Over most of the remainder of the country lie layers of *sedimentary* rock—the limestone of the central plains and much of Connaught, and the sandstone of the Cork and Kerry mountains; these were laid down beneath the shallow seas that once covered this land—the compressed remains of animals and plants, and grains of sand precipitated to the ocean bottom.

Finally, the *metamorphic* rocks are transformations of the other two, changed by intense pressure or by burning heat into richly coloured marbles, slate or quartzite.

Water can have a bizarre and often artistic effect upon this rocky underworld. It works most effectively in limestone, because the rock is vulnerable to acid. Pure water is acidically neutral; rainwater, however, reacts with the carbon dioxide in the atmosphere to form a weak carbonic acid, which has sufficient bite to dissolve considerable quantities of carbonate of lime. It does not need to be 'acid rain'—in the sense in which the term is normally understood—to produce some quite spectacular effects.

Rivers, for example, often disappear when they hit a patch of limestone. The corrosive water may first have enlarged a small crack into a gaping fissure; the gyratory motion of stones and sediments over thousands of years then 'drills' a more or less cylindrical shaft deep down into the rock. These 'swallow-holes' or 'pot-holes' may then continue horizontally as long, winding, subterranean passageways bored out by the acid water. Sometimes the passageways become enlarged to form vast caverns underground.

Limestone itself is suffused with a myriad of tiny cracks and crevices, through which the water from the surface percolates. As the water drips into these caverns, residues of the dissolved limestone that it has carried with it form slender columns of calcite hanging from the roof—the familiar stalactites. And on the floor of the cave where, drop by drop, the water lands incessantly, a similar deposit builds up into the rather thicker stalagmites, rising directly underneath to meet their pendant relatives.

A CATALOGUE OF COMMON CLOUDS

20 *September* 2007 ∿

Recently, on one of my Tuesday forays onto *Today with Pat Kenny*, a listener rang in with what ought to have been a very simple question: which is the most common type of cloud in Irish skies? But it is not an easy question at all—mainly because we in Ireland enjoy, if that is the word, a variety of clouds unmatched perhaps anywhere on earth. Irish skies provide examples of virtually every type of cloud, an eclectic mix which changes from day to day, and even hour by hour.

I suspect the answer to the question may be *stratocumulus*. It often appears as a broken layer or blanket of cloud, typically two to four thousand feet above the ground; usually it is an unthreatening whitish grey in colour, with 'rolls' or undulations, often arranged in long, straight horizontal 'furrows', and is common on dry, cloudy days where there is no particular threat of rain—but not much sun apparent either. Somewhat similar in appearance, but higher up, is *altocumulus*, distinguished from the former mainly by its height—up to eight thousand feet above ground level.

An advancing layer of *altocumulus*, however, is often followed by *altostratus*, which brings a definite threat of rain. It is normally associated with a front, and appears at first as a uniform layer of grey, relatively thin cloud, but thickens rapidly. As the rain begins, the very lowest cloud, the all-too-common *stratus*, is that which you see clinging to the mountain-tops in soft drizzly weather, or scurrying across the sky in ragged patches some hundreds of feet above the ground on a wet and windy day.

The earliest precursors of a spell of frontal rain, however, are the very highest clouds, at 20,000 feet or more above the ground, those feathery filaments of *cirrus* often called 'mares' tails'; they are thin, wispy, fibrous streaks of brilliant white, often exhibiting a silky sheen. *Cirrostratus* is similar in texture, but it forms a continuous sheet, a thin, diaphanous curtain drawn across the sky, often hiding a wan sun attempting to shine faintly through it—an indication that rain is drawing close.

But then in entirely different conditions, when the weather is calm and warm, and the sky relatively clear, surface heating by sunshine produces ascending columns of air, here, there and yonder around the countryside. If the atmosphere is somewhat humid, a cloud forms near the top of each of these invisible pillars—a bulbous rounded cloud with a flat base, resembling a clump of cottonwool or a giant cauliflower.

This is the very common *cumulus* cloud. And when a cumulus cloud becomes very tall, towering perhaps ten or fifteen thousand feet into the sky, meteorologists call it a *cumulonimbus*, clouds which give rise to heavy showers, hail, or even thunderstorms.

WALTER SCOTT'S SIXPENNY WINDS

21 September 2007 ～

The Scottish novelist Sir Walter Scott could be said to have almost single-handedly inspired the modern paradigm of Scottishness. He orchestrated the visit of King George IV to Edinburgh in 1822, and the spectacular pageantry concocted to portray the King as a rather portly reincarnation of Bonnie Prince Charlie made tartans and kilts fashionable and introduced them as symbols of Scottish national identity.

Scott was born in Edinburgh in 1771, the son of a solicitor, and survived a childhood polio in 1773 that left him lame in his right leg for the remainder of his life. He qualified as a lawyer, but it was his poetry, beginning with *The Lay of the Last Minstrel* in 1805, which first brought him fame; then the first of a series of financial crises propelled him into the novel-writing career for which he is best remembered by posterity. He died 175 years ago today on 21 September 1832.

Sir Walter was a keen chronicler of weather matters in his novels. In *The Pirate*, for example, he tells us about a Scandinavian king in olden times called Eric, nicknamed 'Windy Cap', who was able to change the direction of the wind merely by turning his cap around upon his head. He also documents the expertise of one Bessie Miller, who lived at Stromness in the Orkneys around the beginning of the nineteenth century. Bessie, it seems, would sell a favourable wind to any sailor who required one: 'Her fee was extremely moderate being exactly sixpence, for which, as she explained herself, she boiled the kettle, and gave the bark the advantage of her prayers; the wind thus petitioned for was sure to come, she said—although occasionally the mariner had to wait some time for it.'

'The woman's dwelling and appearance', Scott goes on, 'were not unbecoming her pretensions. Her house, which was on the brow of the steep hill on which Stromness is founded, was only accessible by a series of dirty and precipitous lanes, and for exposure might have been the bode of Eolus himself, in whose commodities the inhabitant dealt.' Eolus, you may remember, was the minor Greek deity who, following instructions from the gods, kept the winds imprisoned in a cave on the floating island of Aeolia.

It was Bessie Miller who provided the model for Norna of the Fitful Head—the 'mistress of the potent spell' of Scott's novel who sold favourable winds to mariners. When asked by his tolerant father, 'Why then should Norna not pursue her traffic?' the hero, Mordaunt, sensibly replies: 'Nay I know no reason against it, only I wish she would part with the commodity in smaller quantities. Yesterday she was a wholesale dealer—whoever treated with her had too good a pennyworth.'

WEATHER, WAR AND PEACE

24 *September* 2007 ∿

Lewis Fry Richardson (1881–1953) was born into a well-known Quaker family of the north of England. As his middle name suggests, he was distantly related to certain successful manufacturers of chocolate, and his nephew, Sir Ralph

Richardson, became one of the leading actors of his time. Few may have heard of Lewis Fry Richardson himself, but he is well known to meteorologists as the father of numerical weather prediction, the forecasting of the weather by computerised methods.

Although the concept was not entirely original to Richardson, in *Weather Prediction by Numerical Methods* he described how the process could be carried out—how the future pressure pattern could be calculated if the present state of the atmosphere was accurately known. It was, however, a method without a means; the theory was plausible, but involved an inordinate amount of calculation. Richardson tried it out himself, painstakingly working through the calculations over several months to produce a six-hour forecast, but the result was disappointing. The technique was ignored until the advent of computers, and Richardson died just as his ideas were about to become practicable.

Other theories of Lewis Fry Richardson, however, have yet to be exploited. In keeping with his Quaker origins, Richardson devoted the later part of his life to 'peace research', in which he tried to model mathematically the different national tendencies for war. Delving into the history books, he compiled statistics on all recorded hostilities from 1500 to 1948, classifying them according to the numbers of casualties involved. His highest category was that of wars with a death toll of more than a million people; his lowest was a single casualty—or murder.

Richardson noted, for example, that the tendency for wars to break out during a given period followed a pattern known to scientists as the 'Poisson distribution', a rate of occurrence that applies to many natural phenomena. This, to his mind, indicated a certain mathematical inevitability of conflict, far removed, as he put it, from 'the wide variety of causes that appear in the newspapers every day, including protracted and critical negotiations and the inordinate ambition of the opposing statesmen'. Richardson went on to assign symbols to represent the number of participants, the fatalities, the religious and sociological character-istics of those involved, and then constructed mathematical equations to describe the evolution of hostilities.

Interestingly perhaps, *The Statistics of Deadly Quarrels*, as Richardson called the work, 'dissipated the legend that there are orderly or disorderly peoples; all nations are orderly or disorderly according to the times. There is little hope of forming a group of permanently peace-loving nations to keep the perennially aggressive nations in subjection. Instead the facts support an international order in which a majority of the momentarily peace-loving nations, changing kaleidoscopically in its membership, may hope to restrain a changing minority of momentarily aggressive fellow-nations.'

FROM WEATHER FACTORY TO WEATHER ENGINE

25 September 2007 ~

The eccentric Lewis Fry Richardson, of whom we spoke in this column yesterday, was in a sense the epitome of the nutty professor of the storybooks. One of his best-known scientific papers on turbulence in fluids, for example, unforget-tably begins: 'We have observed the relative motion of a pair of parsnips . . .' Another examines the analogy between sparks and mental images, in an attempt to explain the phenomenon of sudden thoughts. And shortly after the *Titanic* sank, Richardson was observed with his wife blowing a penny whistle in a rowing boat; he was using an umbrella both as an amplifier and a receiver to gauge the strength of the sound reflected from the nearby cliffs, thus anticipating the apparatus that we now call SONAR.

One of Richardson's more far-fetched notions at the time, however, was that a weather forecast could be produced by calculation. His idea was that the future pressure pattern could be calculated if the present state of the atmosphere were accurately known, and clearly it involved an inordinate amount of calculation, but Richardson made a whimsical suggestion: 'Imagine', he wrote, 'a large hall like a theatre, except that the circles and galleries go right through the space usually occupied by the stage. The walls of this chamber are painted to form a map of the globe, and a myriad of persons are at work upon the weather on the part of the map where each sits.' The work, he reckoned, would be supervised by one who would be 'like the conductor of an orchestra in which the instruments are slide-rules and calculating machines'. He called his dream 'the weather factory'.

As it happened, in due course, with the arrival of the electronic computer, Richardson's ideas became a real possibility and, refined over the decades, by the 1960s and '70s they had come to form the basis of modern weather prediction. Every three hours or so, powerful computers were loaded with the latest set of weather observations; they crunched the numbers for an hour or two, and then produced a chart that showed the likely weather pattern for some future time. The predictions increased steadily in accuracy as the models improved over the years, but suffered from the disadvantage that the forecaster received a new chart from the computer only every three hours, or maybe even only every six hours.

To overcome this difficulty, meteorologists now use systems whereby the computer continuously ingests new weather information from ground-based observations and satellite data as they become available, while at the other end the machine makes available, virtually continuously, updated weather charts, based on the very latest situation. The result is a kind of 'continuous weather engine'—the ultimate twenty-first-century realisation, perhaps, of the 'weather factory' envisaged by Lewis Fry Richardson all those years ago.

A CONTROVERSIAL WEATHER PRUSSIAN

27 September 2007 ~

As we continue our trawl of meteorological eccentrics, we might at first innocently suppose H.W. Dove to have been a gentle soul whose name was pronounced exactly as it ought to be. But no! Herr Professor Dr Dove was a larger-than-life, flamboyant character who never encountered controversy without courting it, and whose fellow German meteorologists—unexpectedly to our ears, perhaps—would always refer to him as just 'Hah-Vey Doh-fey'.

Heinrich Wilhelm Dove was born in 1803 in Silesia, then part of Prussia. At the age of 42 he became Professor of Natural Philosophy, the discipline we now call science, at the University of Berlin, and a few years later was appointed Director of the Royal Prussian Institute of Meteorology. There he became for a time what has been described as 'a virtual king of meteorology', his fame being based on painstakingly collecting and analysing climatological observations and drawing from them often controversial, but always plausible, conclusions.

He was one of the first, for example, to relate cloudiness, rain and the general character of the current weather to changes in the atmospheric pressure. Then in his classic and highly respected book *Das Gesetz der Stuerme*, 'The Law of Storms', he developed a concept of the atmosphere based on an equatorial current in apposition to a polar current, with the vicissitudes of the weather in the temperate zones being incidents in the perpetual conflict between these two contrasting regimes. Dove's notion was not too far removed, indeed, from the theory of the polar front developed by the Bergen School of meteorologists several decades later.

Dove had a formidable presence, and a persuasive eloquence sufficient to convince his listeners of the correctness of his notions. But he also held eccentric views, and in promulgating them he made full use of broad generalisations, strange juxta-positions of unusual facts, and plays of intuition which bordered now and then on fantasy. During the 1860s, for example, he objected strenuously to current theories on the thermodynamics of the *foehn* effect, insisting that this was a *moist* wind—and not a dry one, despite the contrary evidence of centuries of popular experience, meteorological theory, and years of reliable measure-ments of the phenomenon. He also insisted, based solely on one of his own charts from which crucial observations had been missing, that the wind blew clockwise, not anticlockwise, around areas of low pressure, quoting the Bible, Aristotle and various literary sources in support of his ideas.

These eccentricities eventually obliged Dove to retire from active meteorological life to work reclusively at his Observatory. But, by the time he died in 1879, his more conventional theories had been enough to earn him after his death an accolade very rare indeed; Dove is one of that very select band of meteorologists to have a crater on the Moon named after him.

ELIZABETH AND THE
MICHAELMAS GOOSE

29 September 2007 ～

S ome might argue that the prize for the most rousing speech in English history should go to Shakespeare's interpretation of Henry v's address to his soldiers on the eve of Agincourt:

> *… he which hath no stomach to this fight,*
> *Let him depart; his passport shall be made,*
> *And crowns for convoy put into his purse …*

But others would say it should go to Elizabeth i for her stirring address to the assembled troops at Tilbury in August 1588, as they prepared to defend her realm against Spain and the Armada. Seated on 'a white horse with its hind quarters dappled iron grey', and with her favourite, Robert Dudley, Earl of Leicester, by her side, Elizabeth proclaimed: 'Let Tyrants fear!'

'At this time I am come amongst you, as you see, not for my recreation and disport, but being resolved in the midst and heat of battle to live and die amongst you all. I know I have the body of a weak and feeble woman, but I have the heart and stomach of a King, and a King of England, too. I myself will take up arms; I myself will be your General, Judge and Rewarder of your virtues in the field.' Of course, Elizabeth did no such thing; almost immediately, she returned by boat to London, but the glowing passion of her words had an electrifying effect upon her audience.

As it happened, it was the weather rather than the English fleet that ultimately defeated the Armada. After a few minor skirmishes, the Spaniards decided that Lady Luck was no longer on their side; they abandoned the campaign and sailed the long way home around the north of Scotland and down the coasts of Donegal and Connaught. But September 1588 was one of the stormiest that century, and many of the Spanish ships were driven onto the rugged Irish coast and wrecked.

Six weeks after her address, Elizabeth seems to have found herself again in Tilbury, and apparently on her return journey this time she supped and stayed the night of Michaelmas, 29 September, with one Sir Neville Umphreyville. History tells us that a roast of goose was placed before the pair that night, and the monarch and Sir Neville both enjoyed it to the full. Towards the banquet's end, the Queen called out for a glass of Burgundy and gave the toast: 'Death to that accursed Armada of the Spaniards!'

Fate responded almost instantaneously. No sooner had Elizabeth spoken than a messenger rushed in to announce the destruction of key vessels of the Spanish fleet in the Blasket Sound on the southwest coast of Ireland, whereupon the Queen proclaimed: 'Henceforth shall a goose commemorate this famous victory!' And it is for this reason, it is said, that goose is the traditional dinner in England on this day.

A DARNED CLOSE-RUN THING —BUT STILL A FAILURE

1 *October* 2007 ⌐

A balloon flight across the Irish Sea ought to be a trivial exercise in aeronautics, bearing in mind the prevailing westerly nature of our winds. The English Channel was crossed in early 1785, for example, less than two years after the very first untethered manned ascent by Pilâtre de Rozier from the Château de la Muette near Paris. But several attempts that same year to cross to Wales from Dublin failed, and it was to be thirty years before the feat was finally accomplished.

Richard Crosbie, a prosperous gentleman from County Wicklow, was the first to try his hand. On a bright morning in May 1785, a large crowd gathered to watch his proposed ascent from what is now Collins Barracks in Dublin. But Crosbie was a man of substance, over 6 feet tall and built to match, and when fully loaded the balloon was unable to lift the heavy aeronaut and his equipment. Rather than disappoint his audience, Crosbie invited the nearest suitable spectator, a lithe young lad called Richard McGwire, to take his place.

With McGwire aboard, the hydrogen-filled balloon was launched and drifted nicely out towards Wales, but it experienced difficulty and ended in the sea near Howth. McGwire was rescued and brought back to a hero's welcome in the city—but that other island was still a hundred miles away.

A Frenchman, Dr Potain, tried the following month. After a short sally in the right direction, Potain drifted south to Wicklow and sank ignominiously to Earth at Roundwood. Then in July, Crosbie went up himself, and after a bumpy start from Leinster

Lawn, he too headed in the right direction—but had to be rescued from the sea mid-channel.

The next endeavour was a quarter of a century later, when one James Sadler, an English aeronaut, came to Dublin specially to try his luck. With great pomp and ceremony he ascended from among the large group of spectators in the grounds of Belvedere House, Drumcondra, 195 years ago today, on 1 October 1812. Drifting gently northeastwards, he made the Isle of Man, and there, changing altitude, he found a northwesterly wind to bring him over Wales. Unfortunately, at the very last moment, another change of wind blew him out to sea again, delivering defeat from the very womb of victory; Sadler was obliged to ditch his balloon, and submit to ignominious rescue by a passing ship.

Five years later, James's son, William Sadler, succeeded. On 22 July 1817, he ascended from Portobello Barracks in a 70-foot balloon, and after a trouble-free and uneventful voyage of six hours he landed in a cornfield two miles from Holyhead on Anglesey. It stood as the only recorded successful landing after a crossing of the Irish Sea by air until the advent of the aeroplane many decades later.

THE FATHER OF GLOBAL WARMING

2 October 2007 ∿

The whole global warming saga might be said to have begun in Leighlinbridge in County Carlow. John Tyndall was born there in August 1820, and he went on to become one of the most distinguished scientists of nineteenth-century Britain. During the 1860s he suggested that slight changes in the composition of the atmosphere might cause variations in the global climate. Tyndall's main interest was water vapour and its impact on solar and terrestrial radiation, but it was he who first identified what we now call the 'greenhouse effect', whether natural or anthropogenic.

But three decades later, during the 1890s, Svante Arrhenius in Sweden became interested in the potential effect on climate of the carbon dioxide resulting from the world's growing industrial activity, and in this context he might be described as the father of our modern theories about global warming. Arrhenius, a chemist by profession, was born in 1859 near Uppsala, Sweden, and in 1891 was appointed a lecturer at Stockholm University. In 1903 he became the first Swede to be awarded a Nobel Prize for Chemistry, and in 1905, upon the founding of the Nobel Institute for Physical Research at Stockholm, he was appointed Rector, a post he held until his retirement in 1927.

Arrhenius painstakingly calculated the amount by which the temperature of the Earth might rise if the amount of carbon dioxide in the atmosphere were doubled and, purely by his own arithmetical dexterity, arrived at a figure of 5°C, impressively close to today's estimates generated by sophisticated climate models and computers. Later in *Världarnas Utveckling* (1906), translated into English as *Worlds in the Making* in 1908, a book directed at a general audience, he suggested that the human emission of CO_2 would be strong enough to prevent the world from entering a new Ice Age, and that a warmer Earth would be needed to feed its rapidly increasing population. He clearly believed at the time that the development of a warmer world would be a change which could be regarded as being for the better.

Strangely enough, Arrhenius's Nobel Prize in 1903 had nothing whatever to do with meteorology or global warming; it concerned his explanation of the fact that neither pure salts nor pure water are good conductors of electricity, but solutions of salts *in* water are very effective in this context. At the time, his Prize was described, somewhat esoterically perhaps, as being 'for extraordinary services rendered to the advancement of chemistry by the electrolytic theory of dissociation'.

OCTOBER ANTICIPATED

3 *October* 2007 ⌒

With the arrival of October, the year has entered the youth of its old age. It marks a turning of the year, being colder, wetter, darker and windier than its predecessors, and there is less sun than in September, if for no other reason than that the days are significantly shorter; the average September day has some four or five hours of sunshine, while the October average is only three.

The temperature on an average October day rises to a mere 13°C or 14°C, 3°C or 4°C less than the September norm. Only very rarely does the temperature exceed 20°C, while at the other end of the scale, ground frost occurs on five or six of October's thirty-one days. Very occasionally, even the air temperature falls below 0°C—an occurrence almost unheard of in September. The waters around our coasts, too, are becoming colder; the usual sea temperature is around 12°C or 13°C, compared to the August peak of 15°C or 16°C. And about every ten years or so, a little snow may fall in October, but it tends to melt as soon as it touches ground, and rarely causes trouble.

But if October is a reminder to us of the rigours of returning winter, the month sometimes has a gentler side: Jack Frost has begun dabbing with his paintbrush. Here and there, against the background of the surviving summer foliage, the first of the autumn leaves are hanging, as the poet Andrew Marvell put it, 'like golden lamps in a green night'. It must be said, however, that Ireland's autumnal splurge, even in areas blessed with an abundance of deciduous trees, can never quite match the razzle-dazzle of Vermont, or of the Black Forest or the Odenwald in Germany.

The colours displayed by the individual trees in autumn are not, *per se*, more splendid or intense in New England or continental Europe than they are in County Wicklow. But the precise trigger for the yearly colour-change varies with the different species, being a combination of declining day-length and the falling temperature. Here in Ireland the autumnal cooling process is gradual and sporadic; some trees that of their nature react early to the falling temperatures will have shed their leaves entirely before other slow-starting species have begun to even think of changing colour.

In places like New England, however, the transition from one season to another is more sudden. The sharp autumnal fall in temperature takes place quickly over a short interval, and catches not just a few trees at a time, but all the trees planning to change their colour. All the leaves are simultaneously transformed, and the whole process from green foliage to bare trees may take little more than two weeks. In between is a short, shrill, spectacular extravaganza of very brilliant colour.

INDEX